CHALLENGES FOR THE CHEMICAL SCIENCES
IN THE 21ST CENTURY

INFORMATION AND COMMUNICATIONS

ORGANIZING COMMITTEE FOR THE WORKSHOP ON
INFORMATION AND COMMUNICATIONS

COMMITTEE ON CHALLENGES FOR THE CHEMICAL SCIENCES
IN THE 21ST CENTURY

BOARD ON CHEMICAL SCIENCES AND TECHNOLOGY

DIVISION ON EARTH AND LIFE STUDIES

NATIONAL RESEARCH COUNCIL
OF THE NATIONAL ACADEMIES

THE NATIONAL ACADEMIES PRESS
Washington, D.C.
www.nap.edu

THE NATIONAL ACADEMIES PRESS • 500 Fifth Street, N.W. • Washington, D.C. 20001

NOTICE: The project that is the subject of this report was approved by the Governing Board of the National Research Council, whose members are drawn from the councils of the National Academy of Sciences, the National Academy of Engineering, and the Institute of Medicine. The members of the committee responsible for the report were chosen for their special competences and with regard for appropriate balance.

Support for this study was provided by the National Research Council, the U.S. Department of Energy (DE-AT-010EE41424, BES DE-FG-02-00ER15040, and DE-AT01-03ER15386), the National Science Foundation (CTS-9908440), the Defense Advanced Research Projects Agency (DOD MDA972-01-M-0001), the U.S. Environmental Protection Agency (R82823301), the American Chemical Society, the American Institute of Chemical Engineers, the Camille and Henry Dreyfus Foundation, Inc. (SG00-093), the National Institute of Standards and Technology (NA1341-01-2-1070 and 43NANB010995), and the National Institutes of Health (NCI-N01-OD-4-2139 and NIGMS-N01-OD-4-2139), and the chemical industry.

All opinions, findings, conclusions, or recommendations expressed herein are those of the authors and do not necessarily reflect the views of the organizations or agencies that provided support for this project.

International Standard Book Number 0-309-08721-X (Book)
International Standard Book Number 0-309-52687-6 (PDF)

Additional copies of this report are available from:

National Academies Press
500 Fifth Street, N.W.
Box 285
Washington, DC 20055
800-624-6242
202-334-3313 (in the Washington Metropolitan Area)
http://www.nap.edu

Copyright 2003 by the National Academy of Sciences. All rights reserved.

Printed in the United States of America

THE NATIONAL ACADEMIES
Advisers to the Nation on Science, Engineering, and Medicine

The **National Academy of Sciences** is a private, nonprofit, self-perpetuating society of distinguished scholars engaged in scientific and engineering research, dedicated to the furtherance of science and technology and to their use for the general welfare. Upon the authority of the charter granted to it by the Congress in 1863, the Academy has a mandate that requires it to advise the federal government on scientific and technical matters. Dr. Bruce M. Alberts is president of the National Academy of Sciences.

The **National Academy of Engineering** was established in 1964, under the charter of the National Academy of Sciences, as a parallel organization of outstanding engineers. It is autonomous in its administration and in the selection of its members, sharing with the National Academy of Sciences the responsibility for advising the federal government. The National Academy of Engineering also sponsors engineering programs aimed at meeting national needs, encourages education and research, and recognizes the superior achievements of engineers. Dr. Wm. A. Wulf is president of the National Academy of Engineering.

The **Institute of Medicine** was established in 1970 by the National Academy of Sciences to secure the services of eminent members of appropriate professions in the examination of policy matters pertaining to the health of the public. The Institute acts under the responsibility given to the National Academy of Sciences by its congressional charter to be an adviser to the federal government and, upon its own initiative, to identify issues of medical care, research, and education. Dr. Harvey V. Fineberg is president of the Institute of Medicine.

The **National Research Council** was organized by the National Academy of Sciences in 1916 to associate the broad community of science and technology with the Academy's purposes of furthering knowledge and advising the federal government. Functioning in accordance with general policies determined by the Academy, the Council has become the principal operating agency of both the National Academy of Sciences and the National Academy of Engineering in providing services to the government, the public, and the scientific and engineering communities. The Council is administered jointly by both Academies and the Institute of Medicine. Dr. Bruce M. Alberts and Dr. Wm. A. Wulf are chair and vice chair, respectively, of the National Research Council.

www.national-academies.org

ORGANIZING COMMITTEE:
WORKSHOP ON INFORMATION AND COMMUNICATIONS

RICHARD C. ALKIRE, University of Illinois, Urbana-Champaign, *Co-Chair*
MARK A. RATNER, Northwestern University, *Co-Chair*
PETER T. CUMMINGS, Vanderbilt University
JUDITH C. HEMPEL, University of Texas, Austin
KENDALL N. HOUK, University of California, Los Angeles
KENNY B. LIPKOWITZ, North Dakota State University
JULIO M. OTTINO, Northwestern University

Liaisons

IGNACIO E. GROSSMANN, Carnegie Mellon University (Steering Committee)
PETER G. WOLYNES, University of California, San Diego (Steering Committee)
SANGTAE KIM, Eli Lilly (BCST)
JOHN C. TULLY, Yale University (BCST)

Staff

JENNIFER J. JACKIW, Program Officer
SYBIL A. PAIGE, Administrative Associate
DOUGLAS J. RABER, Senior Scholar
DAVID C. RASMUSSEN, Program Assistant
ERIC L. SHIPP, Postdoctoral Associate

COMMITTEE ON CHALLENGES FOR THE CHEMICAL SCIENCES IN THE 21ST CENTURY

RONALD BRESLOW, Columbia University, *Co-Chair*
MATTHEW V. TIRRELL, University of California at Santa Barbara, *Co-Chair*
MARK A. BARTEAU, University of Delaware
JACQUELINE K. BARTON, California Institute of Technology
CAROLYN R. BERTOZZI, University of California at Berkeley
ROBERT A. BROWN, Massachusetts Institute of Technology
ALICE P. GAST,[1] Stanford University
IGNACIO E. GROSSMANN, Carnegie Mellon University
JAMES M. MEYER,[2] DuPont Co.
ROYCE W. MURRAY, University of North Carolina at Chapel Hill
PAUL J. REIDER, Amgen, Inc.
WILLIAM R. ROUSH, University of Michigan
MICHAEL L. SHULER, Cornell University
JEFFREY J. SIIROLA, Eastman Chemical Company
GEORGE M. WHITESIDES, Harvard University
PETER G. WOLYNES, University of California, San Diego
RICHARD N. ZARE, Stanford University

Staff

JENNIFER J. JACKIW, Program Officer
CHRISTOPHER K. MURPHY, Program Officer
SYBIL A. PAIGE, Administrative Associate
DOUGLAS J. RABER, Senior Scholar
DAVID C. RASMUSSEN, Program Assistant
ERIC L. SHIPP, Postdoctoral Associate
DOROTHY ZOLANDZ, Director

---

[1] Committee member until July 2001; subsequently Board on Chemical Sciences and Technology (BCST) liaison to the committee in her role as BCST co-chair.
[2] Committee member until March 2002, following his retirement from DuPont.

BOARD ON CHEMICAL SCIENCES AND TECHNOLOGY

ALICE P. GAST, Massachusetts Institute of Technology, *Co-Chair*
WILLIAM KLEMPERER, Harvard University, *Co-Chair*
ARTHUR I. BIENENSTOCK, Stanford University
A. WELFORD CASTLEMAN, JR., The Pennsylvania State University
ANDREA W. CHOW, Caliper Technologies Corp.
THOMAS M. CONNELLY, JR., E. I. du Pont de Nemours & Co.
JEAN DE GRAEVE, Institut de Pathologie, Liège, Belgium
JOSEPH M. DESIMONE, University of North Carolina, Chapel Hill, and North Carolina State University
CATHERINE FENSELAU, University of Maryland
JON FRANKLIN, University of Maryland
MARY L. GOOD, University of Arkansas, Little Rock
RICHARD M. GROSS, Dow Chemical Company
NANCY B. JACKSON, Sandia National Laboratory
SANGTAE KIM, Eli Lilly and Company
THOMAS J. MEYER, Los Alamos National Laboratory
PAUL J. REIDER, Amgen, Inc.
ARNOLD F. STANCELL, Georgia Institute of Technology
ROBERT M. SUSSMAN, Latham & Watkins
JOHN C. TULLY, Yale University
CHI-HUEY WONG, Scripps Research Institute

Staff

JENNIFER J. JACKIW, Program Officer
CHRISTOPHER K. MURPHY, Program Officer
SYBIL A. PAIGE, Administrative Associate
DOUGLAS J. RABER, Senior Scholar
DAVID C. RASMUSSEN, Program Assistant
ERIC L. SHIPP, Postdoctoral Associate
DOROTHY ZOLANDZ, Director

Preface

The Workshop on Information and Communications was held in Washington, D.C., on October 31-November 2, 2002. This was the third in a series of six workshops in the study Challenges for the Chemical Sciences in the 21st Century. The task for each workshop was to address the four themes of discovery, interfaces, challenges, and infrastructure as they relate to the workshop topic (Appendix A).

The Workshop on the Information & Communications brought together a diverse group of participants (Appendix F) from the chemical sciences who were addressed by invited speakers in plenary session on a variety of issues and challenges for the chemical sciences as they relate to computational science and technology. These presentations served as a starting point for discussions and comments by the participants. The participants were then divided into small groups that met periodically during the workshop to further discuss and analyze the relevant issues. Each group provided its discussions to the workshop as a whole.

This report is intended to reflect the concepts discussed and opinions expressed at the Workshop on Information and Communications, and it is not intended to be a comprehensive overview of all of the potential challenges that exist for the chemical sciences in the area of computing. The organizing committee has used this input from workshop participants as a basis for the findings expressed in this report. However, sole responsibility for these findings rests with the organizing committee.

This study was conducted under the auspices of the National Research Council's Board on Chemical Sciences and Technology, with assistance provided by its staff. The committee acknowledges this support.

> Richard C. Alkire and Mark A. Ratner, Co-Chairs,
> Organizing Committee for the Workshop on Information and
>   Communications
> Challenges for the Chemical Sciences in the 21st Century

Acknowledgment of Reviewers

This report has been reviewed in draft form by individuals chosen for their diverse perspectives and technical expertise, in accordance with procedures approved by the National Research Council's (NRC's) Report Review Committee. The purpose of this independent review is to provide candid and critical comments that will assist the institution in making the published report as sound as possible and to ensure that the report meets institutional standards for objectivity, evidence, and responsiveness to the study charge. The review comments and draft manuscript remain confidential to protect the integrity of the deliberative process. We wish to thank the following individuals for their participation in the review of this report:

C. Gordon Bell, Microsoft Bay Area Research Center
Bruce A. Finlayson, University of Washington
Sharon C. Glotzer, University of Michigan
Peter Gund, IBM Life Sciences
Kenneth M. Merz, Jr., The Pennsylvania State University
David H. West, Dow Chemical Company

Although the reviewers listed above have provided many constructive comments and suggestions, they were not asked to endorse the conclusions or recommendations nor did they see the final draft of the report before its release. The review of this report was overseen by Joseph G. Gordon II (IBM Almaden Research Center). Appointed by the National Research Council, he was responsible for making certain that an independent examination of this report was carried out in accordance with institutional procedures and that all review comments were carefully considered. Responsibility for the final content of this report rests entirely with the authoring committee and the institution.

Contents

EXECUTIVE SUMMARY 1
Background and Method, 2
Findings, 3

1 INTRODUCTION: THE HUMAN RESOURCE 7

2 ACCOMPLISHMENTS 12
Major Themes, 12
Some Specific Enabling Accomplishments, 14

3 OPPORTUNITIES, CHALLENGES, AND NEEDS 21
Current Status, 22
Challenges, 23

4 INTERFACES: COOPERATION AND COLLABORATION
ACROSS DISCIPLINES 29
Overarching Themes, 31
    Targeted Design and Open-Ended Discovery, 31
    Flow of Information Between People Within and Among Disciplines, 34
    Multiscale Simulation, 39
    Collaborative Environments, 44
    Education and Training, 47

5 INFRASTRUCTURE: CAPABILITIES AND GOALS 49
Research, 50
Education, 51

Codes, Software, Data and Bandwidth, 53
Anticipated Benefits of Investment in Infrastructure, 55

APPENDIXES
A Statement of Task   63
B Biographies of the Organizing Committee Members   64
C Workshop Agenda   67
D Workshop Presentations,
   Charles H. Bennett, 71
   Anne M. Chaka, 73
   Juan J. de Pablo, 81
   Thom H. Dunning, Jr., 86
   Christodoulos A. Floudas, 116
   Richard Friesner, 125
   James R. Heath, 132
   Dimitrios Maroudas, 133
   Linda R. Petzold, 136
   George C. Schatz, 146
   Larry L. Smarr, 152
   Ellen Stechel, 157
   Dennis J. Underwood, 170
E Biographies of Workshop Speakers   177
F Participants   182
G Reports from Breakout Session Groups   185

Executive Summary

Since publication of the National Research Council (NRC) reports on chemistry in 1985 and chemical engineering in 1988,[1,2] dramatic advances in information technology (IT) have totally changed these communities. During this period, the chemical enterprise and information technology have enjoyed both a remarkably productive and mutually supportive set of advances. These synergies sparked unprecedented growth in the capability and productivity of both fields including the definition of entirely new areas of the chemical enterprise. The chemical enterprise provided information technology with device fabrication processes, new materials, data, models, methods, and (most importantly) people. In turn, information technology provided chemical science and technology with truly remarkable and revolutionary resources for computations, communications, and data management. Indeed, computation has become the strong third component of the chemical science research and development effort, joining experiment and theory.

Sustained mutual growth and interdependence of the chemical and information communities should take account of several unique aspects of the chemical sciences. These include extensive and complex databases that characterize the chemical disciplines; the importance of multiscale simulations that range from molecules to technological processes; the global economic impact of the chemical industry; and the industry's major influence on the nation's health, environment, security, and economic well-being. In planning the future of the chemical

---

[1] *Opportunities in Chemistry,* National Research Council, National Academy Press, Washington, D.C., 1985.
[2]*Frontiers in Chemical Engineering: Research Needs and Opportunities,* National Research Council, National Academy Press, Washington, D.C., 1988.

sciences and technology, it is crucial to recognize the benefits already derived from advances in information technology as well as to point the way to future benefits that will be derived.

BACKGROUND AND METHOD

In October 2002, as part of Challenges for the Chemical Sciences in the 21st Century, the Board on Chemical Sciences and Technology convened a workshop in Washington, D.C., on Information & Communications. The charge to the organizing committee (Appendix A) addressed four specific themes:

- *Discovery:* What major discoveries or advances related to information and communications have been made in the chemical sciences during the last several decades?
- *Interfaces:* What are the major computing-related discoveries and challenges at the interfaces between chemistry/chemical engineering and other disciplines, including biology, environmental science, information science, materials science, and physics?
- *Challenges:* What are the information and communications grand challenges in the chemical sciences and engineering?
- *Infrastructure:* What are the issues at the intersection of computing and the chemical sciences for which there are structural challenges and opportunities—in teaching, research, equipment, codes and software, facilities, and personnel?

The workshop organizing committee assembled a group of top experts to deliver plenary lectures (Appendix C), and recruited an outstanding group of chemical scientists and engineers—from academia, government, national laboratories, and industrial laboratories—to participate in the workshop (Appendix F). Through extensive discussion periods and breakout sessions, the entire group of participants provided valuable input during the course of the workshop. The results of the breakout sessions appear in Appendix G, and written versions of the speakers' presentations are provided in Appendix D. In combination with other references cited in this report, the data collected at the workshop provide the basis for this report.

The structure of the Workshop on Information & Communications followed that of the parent project and each of the other workshops that were held as part of the study of Challenges for the Chemical Sciences in the 21st Century (Materials and Manufacturing, Energy and Transportation, National Security and Homeland Defense, the Environment, and Health and Medicine).

FINDINGS

The information presented in this report enabled the organizing committee to reach a series of conclusions. Advances at the interface between information technology and chemical technology and science are today revolutionizing the way that chemists and engineers carry out their work. Chapter 2 describes accomplishments in the professional development and teaching of people; in methods, models, and databases; and in processes and materials. New tools offered by information technology are fundamentally reshaping research, development, and application activities throughout the chemical sciences. The traditional boundaries between chemistry and chemical engineering are becoming more porous, benefiting both disciplines and facilitating major advances.

Finding: Boundaries between chemistry and chemical engineering are becoming increasingly porous, a positive trend that is greatly facilitated by information technology.
This report contains numerous examples of ways in which databases, computing, and communications play a critical role in catalyzing the integration of chemistry and chemical engineering. The striking pace of this integration has changed the way chemical scientists and engineers do their work, compared to the time of publication of the previous National Research Council reports on chemistry (1985) and chemical engineering (1988).

Finding: Advances in the chemical sciences are enablers for the development of information technology.
Breakthroughs from molecular assembly to interface morphology to process control are at the heart of next-generation IT hardware capabilities. These advances impact computer speed, data storage, network bandwidth, and distributed sensors, among many others. In turn, effective deployment of IT advances within the chemical enterprise will speed discovery of yet more powerful IT engines.

Some of the major challenges to the chemical community can be advanced by IT. Opportunities are plentiful, and challenges and needs remain for further progress. Chapter 3 examines the current status of the research arena in the contexts of computational methodology, training, databases, problem solving, optimization, communications capabilities, and supply-chain modeling. The challenges are then described that will certainly arise, as computing capabilities and information technology continue to grow, and the modeling tasks for the chemical community become more complex.

Finding: There are major societal and civic problems that challenge the chemical community. These problems should be addressed by chemistry and chemical engineering, aided by IT advances.
These societal issues include providing stewardship of the land, contributing

to the betterment of human health and physical welfare, ensuring an informed citizenry through education, facilitating more thoughtful and informed decision making, and protecting and securing the society.

Finding: The nation's technological and economic progress can be advanced by addressing critical needs and opportunities within the chemical sciences through use of new and improved information technology tools.
Bringing the power of IT advances to bear will greatly enhance both targeted design through multidisciplinary team efforts and decentralized curiosity-driven research of individual investigators. Both approaches are important, but they will depend upon IT resources in different ways.

Finding: To sustain advances in chemical science and technology, new approaches and IT infrastructures are needed for the development, support, and management of computer codes and databases.
Significant breakthroughs are needed to provide new means to deal with complex systems on a rational basis, to integrate simulations with theory and experiment, and to construct multi-scale simulations of entire systems.

Pervasive computing and data management will complement and aid the role of intuition in allowing science and engineering to take full advantage of human resources. Chapter 4 addresses the ways in which information technology and computation can provide new capabilities for cooperation and collaboration across disciplines. Training chemical scientists and engineers to take strategic advantage of advances in information technology will be of particular importance. Overarching themes such as targeted design, curiosity-driven research, flow of information, multiscale simulation, and collaborative environments will become increasingly significant as information technology becomes yet more capable and the chemical community undertakes even more intricate problems.

Finding: Computation and information technology provide a key enabling force for lowering barriers among the disciplines that comprise the chemical enterprise and closely related fields.
Identification of mutual interests among disciplines and removal of the barriers to successful communication among constituencies are essential for increasing the overall effectiveness of the system. The processes of identification and removal are still in their infancy.

Finding: Addressing critical challenges at the interfaces with other scientific and engineering disciplines will enable chemistry and chemical engineering to contribute even more effectively to the nation's technological and economic progress.

The most important challenge involves people. Advances in IT that facilitate self-organization of problem-solving groups with common interests across disciplinary boundaries will impact strongly both understanding-based and application-driven projects. The essential resource driving the interface of IT and the chemical sciences is human ingenuity.

Finding: The capability to explore in the virtual world will enable society to become better educated and informed about the chemical sciences. Conveying the intellectual depth, centrality, societal benefits, and creative challenges of molecular systems will be greatly facilitated by the use of modeling, visualization, data manipulation, and real-time responses. All of these new capabilities will provide unparalleled environments for learning, understanding, and creating new knowledge.

Finding: The growing dependence of the chemical enterprise on use of information technology requires that chemical professionals have extensive education and training in modern IT methods.
This training should include data structures, software design, and graphics. Because data and its use comprise such important aspects of chemistry and chemical engineering, and because appropriate use of IT resources can empower unprecedented advances in the chemical arena, it is crucial that the appropriate training, at all levels, be a part of chemical education.

Looking to the future, we need to build upon these advances to enable computational discovery and computational design to become standard components of broad education and training goals in our society. In this way, the human resources will be available to create, as well as to realize and embrace, the capabilities, challenges, and opportunities provided by the chemical sciences through advanced information technology. Chapter 5 deals with capabilities and goals—structural challenges and opportunities in the areas of research, teaching, codes, software, data and bandwidth. Major issues of infrastructure must be addressed if the nation is to maintain and improve the remarkable economic productivity, scientific advances, and societal importance of the chemical sciences and technology.

Finding: Federal research support for individual investigators and for curiosity-driven research is crucial for advances in basic theory, formalisms, methods, applications, and understanding.
History shows that the investment in long-term, high-risk research in the chemical sciences must be maintained to ensure continued R&D progress that provides the nation's technological and economic well-being. Large-scale, large-group efforts are complementary to individual investigator projects—both are crucial, and both are critically dependent on next-generation IT infrastructure.

Finding: A strong infrastructure at the intersection with information technology will be critical for the success of the nation's research investment in chemical science and technology.
The infrastructure includes hardware, computing facilities, research support, communications links, and educational structures. Infrastructure enhancements will provide substantial advantages in the pursuit of teaching, research, and development. Chemists and chemical engineers will need to be ready to take full advantage of capabilities that are increasing exponentially.

RECOMMENDATIONS

These findings show that the intersection of chemistry and chemical engineering with computing and information technology represents a frontier ripe with opportunity. Major technical progress can be expected only if additional resources are provided for research, education, and infrastructure. While this report identifies many needs and opportunities, the path forward is not yet fully defined and will require additional analysis.

Recommendation: Federal agencies, in cooperation with the chemical sciences and information technology communities will need to carry out a comprehensive assessment of the chemical sciences–information technology infrastructure portfolio.
The information provided by such an assessment will provide federal funding agencies with a sound basis for planning their future investments in both disciplinary and cross-disciplinary research.

Recommendation. In order to take full advantage of the emerging Grid-based IT infrastructure, federal agencies—in cooperation with the chemical sciences and information technology communities—should consider establishing several collaborative data–modeling environments.
By integrating software, interpretation, data, visualization, networking, and commodity computing, and using web services to ensure universal access, these collaborative environments could impact tremendously the value of IT for the chemical community. They are ideal structures for distributed learning, research, insight, and development on major issues confronting both the chemical community and the larger society.

1

Introduction: The Human Resource

Information technology is transforming American and global society. The availability of deep information resources provides a fundamentally new capability for understanding, storing, developing, and integrating information. Information and communication tools in chemical science and technology have already provided an unprecedented capability for modeling molecular structures and processes, a capability that has contributed to fundamental new understanding as well as new technological products based on the physical and life sciences.

Chemistry and chemical engineering are being transformed by the availability of information technology, modeling capabilities, and computational power. The chemical sciences in the twenty-first century will include information, computation, and communications capabilities as both assets and challenges. The assets are clear in terms of what we already can accomplish: we can model many systems with accuracy comparable to or exceeding that of experiment; we can rapidly and effectively extend theoretical conceptual development toward modeling capabilities; and we can store, retrieve, integrate, and display information effectively and helpfully.

The challenges come at several levels. Major exploration will be needed to develop new and better tools, educational techniques, computational and modeling strategies, and integrative approaches. The exploration will create demands in two areas: *chemical information technology* and the *people who will do the work.*

The two traditional components of the scientific method, observation and hypothesis, have led to a formidable array of experimental and theoretical tools. Since the end of World War II, computation and modeling has advanced to become the strong third component, one that can integrate experiment and theory with application. Advances in information technology (IT) in communications, modeling, and computing have substantially increased the capabilities of chemistry and chemical engineering. Effectively harnessing new and future IT advances will present a great challenge for chemical science, but success will provide both contributions to fundamental knowledge and benefits to our society in health, welfare, and security.

Looking to the future, we need to build upon these advances to enable computational discovery and computational design to become standard components of broad education and training goals in our society. In this way, the human resources will be available to create, as well as to realize and embrace, the capabilities, challenges, and opportunities provided by the chemical sciences through advanced information technology.

Chemists and chemical engineers, and the processes and goods that they produce, have probably the largest impact of any science/engineering discipline on our economy and on our environment. The chemical industry employs over one million workers in the United States and indirectly generates an additional five million jobs; this business of chemistry contributes nearly $20 billion annually to federal, state, and local tax revenues.[1] Investment in chemical R&D is estimated to provide an annual return of 17% after taxes.[2] Chemical manufacturing (including chemicals, allied products, petroleum, coal products, rubber, and plastics) produces 1.9% of the U.S. gross domestic product (GDP) and approximately 17% of the for the manufacturing sector.[3] The chemical industry supplies nearly $1 out of every $10 of U.S. exports,[4] and in 2002 its total exports of $82 billion ranked second among exporting sectors.[5]

It is therefore especially important that we, as a society, take steps to assure that the chemical enterprise maintain its cutting-edge capability in teaching, research, development, and production. It is also important that the chemical enterprise provide leadership in economic growth and environmental quality. All of these goals require increased capability for chemists and chemical engineers to utilize, efficiently and creatively, the capabilities offered by information technology.

Advances in the chemical sciences enabled major achievements in medicine, life science, earth science, physics, engineering, and environmental science. These advances in the productivity, quality of life, security, and economic vitality of our society flowed directly from the efforts of people who work in those fields. How

---

[1]*Guide to the Business of Chemistry,* American Chemistry Council, Arlington, VA, 2002; http://www.accnewsmedia.com/docs/300/292.pdf.

[2]*Measuring Up: Research & Development Counts in the Chemical Industry,* Council for Chemical Research, Washington, D.C., 2000; http://www.ccrhq.org/news/studyindex.html.

[3]U.S. Department of Commerce, Bureau of Economic Analysis, *Industry Accounts Data, Gross domestic product by industry:* http://www.bea.doc.gov/bea/dn2/gposhr.htm.

[4]U.S. Department of Commerce, Technology Administration: *The Chemical Industry:* http://www.technology.gov/reports.htm.

[5]*Chemical & Engineering News* **2003**, *81(27),* 64.

will we as a community utilize the remarkable capabilities provided by IT to teach, train, inspire, challenge, and reward not only the professionals within our discipline but also those in allied fields whose work depends on understanding and using concepts and ideas from the chemical sciences?

This report is written by the committee that organized a workshop held in Washington, D.C., in October 2002, to address ways in which chemists and chemical engineers could focus their R&D efforts on the solution of problems related to computing and information technology. A series of speakers (Appendix E) presented lectures (Appendix D) on topics that covered different aspects of the problem, and they addressed issues in all areas of chemical science and engineering. Considerable input for the report was also provided by a series of breakout sessions (Appendix G) in which all workshop attendees participated (Appendix F). These breakout sessions explored the ways in which chemists and chemical engineers already have contributed to solving computationally related problems, the technical challenges that they can help to overcome in the future, and the barriers that will have to be overcome for them to do so. The questions addressed in the four breakout sessions were:

- *Discovery:* What major discoveries or advances related to information and communications have been made in the chemical sciences during the last several decades?
- *Interfaces:* What are the major computing-related discoveries and challenges at the interfaces between chemistry–chemical engineering and other disciplines, including biology, environmental science, information science, materials science, and physics?
- *Challenges:* What are the information and communications grand challenges in the chemical sciences and engineering?
- *Infrastructure*: What are the issues at the intersection of computing and communications with the chemical sciences for which there are structural challenges and opportunities—in teaching, research, equipment, codes and software, facilities, and personnel?

The world of computing has grown at an extraordinary pace in the last half century.[6] During the early stages, the impact of this growth was restricted to a small segment of the population, even within the technical community. But as the expanded power of computer technology made it possible to undertake significant new areas of research, the technical community began to embrace this new technology more broadly. Perhaps the seminal event in changing the culture was

---

[6]For example, "Moore's Law," originally stated as "The complexity for minimum component costs has increased at a rate of roughly a factor of two per year," Moore, G. E., *Electronics* **1965**, *38 (8)* 114-117. This has been restated as "Moore's Law, the doubling of transistors every couple of years"; (*http://www.intel.com/research/silicon/mooreslaw.htm*).

the introduction of personal computers in the 1980s. By 1994 the number of U.S. households with personal computers had reached 24%,[7] and this increased to 54% by 1999.[8] For Japan, the analogous numbers are 12% in 1986, 49% in 1999, and 88% in 2002.[9]

The key to the future is the human resource.[10] Computers are extraordinarily powerful tools, but they do only what people tell them to do. There is a remarkable synergy between humans and computers, because high levels of human creativity are needed to push the capabilities of computers in solving research problems. At the same time, computers have enabled an astonishing increase in human creativity, allowing us to undertake problems that previously were far too complex or too time-consuming to even consider. Our technical future is strongly linked to our ability to take maximum advantage of the computer as a way of doing routine tasks more rapidly, beginning to undertake tasks that we could not do before, and facilitating the creativity of the human mind in ways that we have not yet imagined.

Like so many other aspects of the information technology universe, the use of computational resources for addressing chemical systems has been growing rapidly. Advances in experiment and theory, the other two principal research and development modes in chemical science, have also developed rapidly. The advances in the chemical sciences enabled by exponential growth of computational capability, data storage, and communication bandwidth are by far the most striking and profound change in the past two decades. This remarkable growth has been stressed elsewhere,[11,12] and is clearly stated by Jack Dongarra, one of the world's foremost experts in scientific computing, who has argued that

> ...the rising tide resulting from advances in information technology shows no respect for established order. Those who are unwilling to adapt in response to this profound movement not only lose access to the opportunities that the infor-

---

[7]National Telecommunications and Information Administration, *Falling Through the Net, Toward Digital Inclusion,* 2000, *http://www.ntia.doc.gov/ntiahome/digitaldivide/.*

[8]Arbitron, *Pathfinder Study,* 1999, New York, *http://internet.arbitron.com/main1.htm.*

[9]*http://www.jinjapan.org/stat/stats/10LIV43.html.*

[10]*Beyond Productivity: Information, Technology, Innovation, and Creativity,* Mitchell, W. J.; Inouye, A. S.; Blumenthal, M. S., Eds. National Research Council, The National Academies Press, Washington, DC, 2003.

[11]*Revolutionizing Science and Engineering through Cyber-infrastructure,* Report of the National Science Foundation Blue-Ribbon Advisory Panel on Cyberinfrastructure, Alliance for Community Technology, Ann Arbor, MI, 2003 (the Atkins committee report); *http://www.communitytechnology. org/nsf_ci_report/.*

[12]*Science and Engineering Infrastructure for the 21st Century: The role of the National Science Foundation,* National Science Board, Arlington, VA, 2003; This report lists as one of its key recommendations to "Develop and deploy an advanced cyberinfrastructure to enable new S&E in the 21st century."

mation technology revolution is creating, they risk being rendered obsolete by smarter, more agile or more daring competitors.[13]

At the current rate of change, communications and computing capabilities will increase tenfold every five years. Such rapid increase of capability means that some problems that are unsolvable today will be straightforward in five years. The societal implications are powerful. To deal with these assets, opportunities, and challenges will require both an awareness of the promise and a commitment of financial and human resources to take advantage of the truly revolutionary advances that information technology offers to the world of chemical science.

---

[13] See T. Dunning, Appendix D.

2

Accomplishments

The chemical sciences begin the twenty-first century with an enviable record of accomplishment in education, research, technology development, and societal impact. The record also extends to information technology (IT), where it works in both directions—the chemical sciences are impacted by, as well as exert impact upon, information technology.

MAJOR THEMES

The chemical sciences and information technology form a mutually supportive partnership. This dates to the early years when IT was still in the vacuum-tube age. The chemical sciences have provided construction modalities for computers, ranging from polyimide packaging to organic photoresists and chemical vapor deposition. Today chemical sciences contribute to the partnership in three major areas. The chemical sciences provide

1. people who develop and extend information technology through expertise that ranges from algorithm development to programming, from process and materials research through database and graphics development, and from display technology research to mobile power sources;

2. theoretical models and methods on which software programs can be based, along with a huge amount of significant, wide-ranging, and unique data to construct crucial databases; and

3. processes and materials for construction of information networks and computers—a microelectronic fabrication facility involves many processing operations that are familiar to chemical engineers.

In turn, information technology has provided for the chemical sciences a series of major enablers of processes and schemes for enriching both what chemists and chemical engineers can do and the ease with which they can do it.

- *Languages and Representations:* In addition to the traditional upper level programming languages, new approaches—including scripting languages, markup languages, and structural representations—have greatly facilitated the successful use of IT within the chemical sciences. These languages make it possible to use distributed computers very efficiently. This can be useful in exploring multidimensional design, structure, and dynamics problems.
- *Informatics and Databases:* Starting with the remarkably complete and powerful informational database represented by *Chemical Abstracts*,[1] we have witnessed in the last 20 years a striking development of databases, database relationships, data-mining tools and tutorial structures. These enable the kind of research to be done in the chemical sciences that could never have been done before. From something as straightforward as the Protein Data Bank (PDB)[2] to the most subtle data-mining algorithms, information technology has permitted all scientists to use the huge resource of chemical data in a highly interactive, reaonably effective fashion. Other contributions include the developments of string representations for chemical structures.
- *Computing Capability, Integration, and Access:* These have enabled cutting-edge work in the chemical sciences that resulted in a number of Nobel Prizes in Chemistry, National Medals of Technology, and National Medals of Science. More broadly, they have enabled research modalities within the chemical sciences that could not have been accomplished before the advent of computers. Much of this report is devoted to computational modeling and numerical theory—areas of research and development that didn't exist before 1950. Striking examples such as quantum chemical calculations of molecular electronic structure, Monte Carlo calculations of equations of state for gases and liquids, or molecular dynamics simulations of the structures of high-temperature and high-pressure phases represent major extensions of the traditional concepts of chemistry. Integrated models for plant design and control, Monte Carlo models for mixtures, polymer structure and dynamics, and quantum and classical dynamics models of reaction and diffusion systems provide chemical engineers with an ability to predict the properties of complex systems that, once again, was simply unobtainable before 1950.
- *Bandwidth and Communication Capabilities:* These have enabled new levels of collaboration for chemical scientists and engineers, who now have in-

---

[1] A product of the American Chemical Society, *http://www.cas.org/*.

[2] The PDB is a repository for three-dimensional biological macromolecular structure data; *http://www.rcsb.org/pdb/*; Berman, H. M.; Westbrook, Feng, Z.; Gilliland, G.; Bhat, T. N.; Weissig, H.; Shindyalov, I. N.; Bourne, P. E. *Nucleic Acids Research* **2000,** *28,* 235-242.

stantaneous access to information, to measurements, and to the control of entire systems. The impact of increased bandwidth typically has received less attention than that of advances in informatics and computing. However, it is key to much of the future societal and educational development of the chemical sciences—to the processes that will allow chemists and chemical engineers to interact and collaborate with one another, with other scientists, with industrial and medical practitioners, and with users that require chemical information, methods, and models. Similarly, increases in network and memory bus speed have made computation a more powerful tool for modeling in chemical science.

The emphasis on Moore's Law as a measure of steady increase in computing power is well known. Equally remarkable growth has also occurred in algorithm development. Figure 2-1 shows performance increases over the past three decades derived from computation methods as well as from supercomputer hardware, noting approximate dates when improvements were introduced. It may be recognized that growth in algorithm speed and reliability has had a significant impact on the emergence of software tools for the development and integration of complex software systems and the visualization of results.

SOME SPECIFIC ENABLING ACCOMPLISHMENTS

The chemical sciences have heavily influenced the field of information and communications during the past five decades. Examples are so common that we now take them for granted or even dismiss them as simply being a part of modern society. But in retrospect it is clear that scientists, engineers and technologists have used training in the chemical sciences to contribute significantly to information technology and its use. Chemical scientists have built methods, models, and databases to take advantage of IT capabilities, and they have explored and developed materials from knowledge based at the molecular level. Some examples follow.

1. People:

Professionals with backgrounds in chemical science have provided a major intellectual resource in the industrial, government, and academic sectors of society. Such chemical problems as process design, optimization of photorefractive polymers, or organization and searching of massive databases are precisely the sort of complicated and demanding environment to provide excellent training for work in information technology policy and development.

2. Methods, Models, and Databases

• *Modeling:* A series of model chemistries has been developed and partially completed. This has facilitated the use of computational techniques—in a relatively straightforward way—for specific problems such as molecular

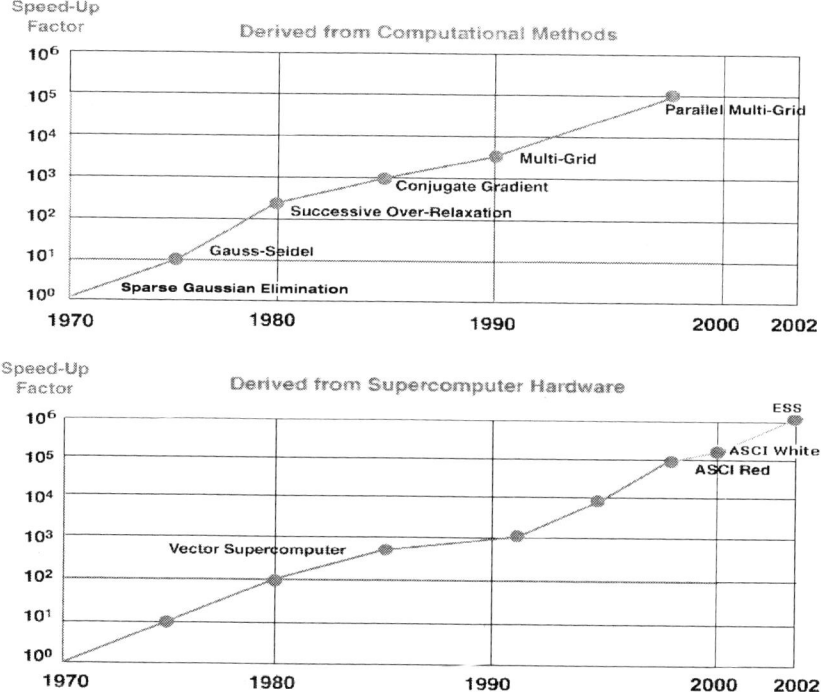

FIGURE 2-1 Speedup resulting from software and hardware developments. Updated from charts in *Grand Challenges: High Performance Computing and Communications*, Office of Science and Technology Policy Committee on Physical, Mathematical and Engineering Sciences, 1992; SIAM Working Group on CSE Education, *SIAM Rev.* **2001**, *43:1*, 163-177; see also L. Petzold, Appendix D.

structure. The availability of quantum chemical codes has made it possible to solve, with high accuracy and precision, many problems associated with the ground state for organic molecules. This has been used extensively to solve problems spanning the range from molecules in outer space to drug design.

• *Multiscale Computational Integration:* The beginnings of multiscale understanding and methodologies are being developed. These include hierarchical approaches such as using molecular dynamics to compute diffusion coefficients or materials moduli that in turn can allow extended scale descriptions of real materials, or using quantum chemistry to define electrical and optical susceptibility that can then be used in full materials modeling. Such efforts mark the beginnings of multiscale computational integration.

• *Integrated Data Management:* Computational bandwidth and extant databases are being utilized to develop user-friendly and integrated data management and interpretation tools. Such utilities as Web of Science, Chemical

Abstracts Online, and the structural libraries developed at Cambridge,[3] Brookhaven,[4] and Kyoto[5] are examples of using information technology combined with chemical data to facilitate research and understanding at all levels.

3. Processes and Materials

- *Optical Fibers:* The development of optical fibers—prepared from silica glass purified by removal of bubbles and moisture and capable of transmitting light over great distances—has made possible the communication backbone for the early twenty-first century.

- *Photoresist Technology:* The development of photoresist technology, based on polymer chemistry, has been an integral part of chip, packaging, and circuit board manufacturing for the past four decades.

- *Copper Electrochemical Technology:* The manufacture of complex microstructures for on-chip interconnects requires multiple layers of metallization. Copper electrochemical technology was introduced by IBM in 1999 and is now used widely as a basic chip fabrication process. The process depends critically on the action of solution additives that influence growth patterns during electrodeposition.

- *Magnetic Films:* Magnetic data storage is a $60 billion industry that is based on the use of thin film heads fabricated by electrochemical methods. Since their introduction in 1979, steady improvements in storage technology have decreased the cost of storage from $200/Mbyte to about $0.001/Mbyte today.[6]

Examples such as the preceding demonstrate that the remarkable IT advances we have seen—in speed, data storage, and communication bandwidth—have been facilitated in no small way by contributions from the chemical sciences. A recent article by Theis and Horn, of IBM, discusses basic research in the information technology industry and describes the ongoing—indeed growing—importance of nanoscale chemistry in the IT community.[7]

---

[3]The Cambridge Structural Database (CSD) is a repository for crystal structure information for organic and metal-organic compounds analyzed by X-ray or neutron diffraction techniques, *http://www.ccdc.cam.ac.uk/prods/csd/csd.html.*

[4]The PDB; see footnote 3.

[5]The Kyoto Encyclopedia of Genes and Genomes (KEGG) is a bioinformatics resource for genome information, *http://www.genome.ad.jp/kegg/.*

[6]Romankiw, L.T. *J. Mag. Soc. Japan* **2000**, *24:1.*

[7]Theis, T. N.; Horn, P. M. *Physics Today* **2003**, *56(7); http://www.physicstoday.org/vol-56/iss-7/p44.html.*

Finding: Advances in the chemical sciences are enablers for the development of information technology.
Breakthroughs from molecular assembly to interface morphology to process control are at the heart of next-generation IT hardware capabilities. These advances impact computer speed, data storage, network bandwidth, and distributed sensors, among many others. In turn, effective deployment of IT advances within the chemical enterprise will speed discovery of yet more powerful IT engines.

The flow of information and influence goes in both directions; just as chemistry and chemical engineering have had a significant influence on the development of computing, information technology has helped to produce major advances in chemical science and engineering. Examples include the following:

- *Computer-Aided Drug Design:* This has become a significant contributor in the discovery and development of new pharmaceuticals. Some examples of molecules created with the direct input of computational chemistry include the antibacterial agent norfloxacin, the glaucoma drug dorzolamide hydrochloride marketed as Trusopt, indinavir sulfate marked as Crixivan, the protease inhibitor for AIDS, marketed as Norvir, the herbicides metamitron and bromobutide, and the agrochemical fungicide myclobutanil.
- *Simulation and Computational Methods for Design and Operation*: These techniques are used for something as simple as a new colorant or flavor species to something as complicated as the integrated design and real-time optimization of a chemical plant or a suicide enzyme inhibitor. Predictive capabilities are now good enough to obtain phase diagrams for real gases with accuracies exceeding most experiments. Simulations are now beginning to address more complex systems, including polymers, biomolecules, and self-assembling systems. Molecular dynamics (MD) simulation of billions of atoms is now possible, permitting both understanding and prediction of such phenomena as phase and fracture behavior.
- *Integrated Control and Monitoring Systems*: This approach recently permitted the IBM Corporation to dedicate a multibillion-dollar semiconductor fabrication line in East Fishkill, New York, that is controlled almost entirely by computers and employs remarkably few people to maintain production.
- *Chemoinformatics:* A new area of chemical science has begun, one that often is called chemoinformatics (one negative aspect of information technology is the ugliness of some of the new words that it has engendered). Chemoinformatics is defined as the "application of computational techniques to the discovery, management, interpretation and manipulation of chemical information and data."[8] It is closely related to the growth of such techniques as high-

---

[8]*Naturejobs* **2002**, *419*, 4-7.

throughput screening and combinational chemistry, and is a strong growth area both in employment and in its ability to use fundamental knowledge to yield practical advances.
- *Molecular Electronic Structure Calculations*: This capability has become available to the community in a number of integrated computer codes. These make it possible for chemists, engineers, physicists, astronomers, geologists, and educators to model the structure and properties of a given molecule (within limits) to high accuracy. This ability has led to a new way of doing science, based not on artificial models but on accurate quantum calculations of actual chemical species. In some cases, accurate calculations can replace experiments that are expensive or dangerous or involve animals.

In a curious way, some of the important accomplishments in chemical science with respect to information technology involve realization of strengths and definition in other areas—fields that can (or perhaps must) in the future take advantage of exponential advances in IT implied by Moore's Law.[9] These accomplishments address fundamental issues or enabling methods to solve major problems, often outside the chemical sciences, as illustrated by the following examples:

- The chemical industry does a far better job than either universities or government laboratories of integrating capabilities and training across disciplines and backgrounds. The ability to integrate expertise is absolutely crucial to the success of modeling efforts in the chemical, pharmaceutical, and energy sectors.
- Areas that have data streams rich enough to require truly massive data management capability have undergone major development. These areas include combinatorial chemistry, the use of microfluidics to study catalysts, and the rapidly expanding capability to label enormous numbers of individual objects, such as cans of soup or shipping parcels—using, for example, inexpensive polymer-based transistor identification tags. Remarkable advances in security, economy, and environmental assurance can follow from computational monitoring, modeling, and communicating in real time the flow of these material objects within society.
- Supply-chain optimization and process optimization are showing initial success. Full utilization of such capabilities can have significant positive impact on the competitiveness and capability of industrial production, homeland defense and security, and the quality of the environment.
- Integrated modeling efforts are proving highly valuable to industry in such areas as drug design and properties control. Richard M. Gross, vice president and

---

[9]Moore, G.E., *Electronics* **1965**, *38 (8)* 114-117; *http://www.intel.com/research/silicon/ mooreslaw.htm*.

director of research and development for the Dow Chemical Company, described at another workshop[10] the critically important role of computational chemistry in designing the low-k (low dielectric constant) resin, SiLK. The dielectric, mechanical, and thermal properties of a large group of polymers were predicted computationally in order to identify a much narrower group of targets on which synthetic efforts were focused. Close integration of computation and chemistry during very early stages of discovery and process invention were key for getting sample material into the marketplace within six months of the original go-ahead for pursuing the idea.

In many areas, simulation capabilities now make it possible to go beyond simplistic models to truly integrated, high-accuracy simulations that can provide an accurate guide to the properties of actual structures. Simulations combine the scientist's search for truth with the engineer's desire for targeted design, the deterministic solution of well-defined sets of equations with the use and understanding of stochastic and probabilistic arguments, and the targeted strengths of the research institute with the multilevel demands of R&D business. Sophisticated simulation tools that combine all of these aspects are beginning to be developed, and they constitute a major advance and opportunity for chemical science.

Perhaps the most significant accomplishment of all is the fundamental reshaping of the teaching, learning, and research and development activities that are likely to be carried out in the chemical sciences by taking strategic advantage of new information technology tools. Within chemical engineering and chemistry, we are approaching an era of "pervasive computing." In this picture, computation and information will be universal in the classroom, the laboratory, and manufacturing areas. Already, organic chemists use databases and data mining to suggest molecular structures, quantum chemistry to predict their stability, and statistical mechanics methods (Monte Carlo, molecular dynamics) to calculate their properties and interactions with other species. What once was the esoteric domain of the theoretical chemist now encompasses scientists and engineers from high schools to seasoned professionals. This integration of modeling, simulation, and data across many sectors of the society is just beginning, but it is already a major strength and accomplishment.

Finding: Boundaries between chemistry and chemical engineering are becoming increasingly porous, a positive trend that is greatly facilitated by information technology.

This report contains numerous examples of ways in which databases, computing, and communications play a critical role in catalyzing the integration

---

[10]*Reducing the Time from Basic Research to Innovation in the Chemical Sciences, A Workshop Report to the Chemical Sciences Roundtable,* National Research Council, The National Academies Press, Washington, D.C., 2003.

of chemistry and chemical engineering. The striking pace of this integration has changed the way chemical scientists and engineers do their work, compared to the time of publication of the previous National Research Council reports on chemistry[11] (1985) and chemical engineering[12] (1988).

Among the many examples of this trend is the study on Challenges for the Chemical Sciences in the 21st Century. The integration of chemistry and chemical engineering is a common thread that runs throughout the report *Beyond the Molecular Frontier: Challenges for Chemistry and Chemical Engineering*,[13] as well as the six accompanying reports on societal needs (of which this report is one).[14,15,16,17,18] These describe both the progress and the future needs for increasing cooperation that will link chemical science and chemical technology.

Many of the barriers and interfaces are discussed in this report, but further analysis and action will be needed on many fronts—by individual scientists and engineers and by administrators and decision makers—in universities and individual departments, in companies and federal laboratories, and in those agencies that provide financial support for the nation's research investment in the chemical sciences.

---

[11]*Opportunities in Chemistry,* National Research Council, National Academy Press, Washington, D.C., 1985.

[12]*Frontiers in Chemical Engineering: Research Needs and Opportunities,* National Research Council, National Academy Press, Washington, D.C., 1988.

[13]*Beyond the Molecular Frontier: Challenges for Chemistry and Chemical Engineering,* National Research Council, The National Academies Press, Washington, D.C., 2003.

[14]*Challenges for the Chemical Sciences in the 21st Century: National Security & Homeland Defense,* National Research Council, The National Academies Press, Washington, D.C., 2002.

[15]*Challenges for the Chemical Sciences in the 21st Century: Materials Science and Technology,* National Research Council, The National Academies Press, Washington, D.C., 2003.

[16]*Challenges for the Chemical Sciences in the 21st Century: Energy and Transportation,* National Research Council, The National Academies Press, Washington, D.C., 2003 (in preparation).

[17]*Challenges for the Chemical Sciences in the 21st Century: The Environment,* National Research Council, The National Academies Press, Washington, D.C., 2003 ..

[18]*Challenges for the Chemical Sciences in the 21st Century: Health and Medicine,* National Research Council, The National Academies Press, Washington, D.C., 2003 (in preparation).

3

Opportunities, Challenges, and Needs

Information technology (IT) is a major enabler for the chemical sciences. It has provided the chemical scientist with powerful computers, extensive database structures, and wide-bandwidth communication. It permits imagination, envisioning, information integration and probing, design at all levels, and communication and education modalities of an entirely new kind.

The richness and capability of effective data management and data sharing already permit, and in the future will facilitate even more successfully, entirely new kinds of understanding. Combining modeling with integrated data may permit the community to predict with higher reliability issues of risk, environmental impact, and the projected behavior of molecules throughout their entire life cycle in the environment. Major progress on issues of societal policy—ranging from energy to manufacturability, from economic viability to environmental impact, and from sustainable development to responsible care to optimal use of matter and materials—might all follow from the integrated capabilities for data handling and system modeling provided by advances in information technology.

This field is beginning to see the development of cooperative environments, where learning, research, development, and design are carried out utilizing both modeling and data access: this cooperative environment for understanding may be the most significant next contribution of IT within the chemical sciences.

The questions posed by chemists and chemical engineers are often too complex to be solved quantitatively. Nevertheless, the underlying physical laws provide a framework for thinking that, together with empirical measurement, has allowed researchers to develop an intuition about complex behavior. Some of these complicated problems will yield to quantitative solution as computational power continues to increase. Most will not, however, at least in the foreseeable future. These will continue to be addressed by intuition and experiment. The data-handling and visualization capabilities of modern computing will increasingly become an essential aid in developing intuition, simple models, and underlying physical pictures. Thus, information technology has changed and will continue to change the way we think about chemical problems, thereby opening up new vistas well beyond the problems that can be addressed directly by large-scale computation.

CURRENT STATUS

Computational modeling currently excels in the highly accurate computation of small structures and some gas-phase properties, where experiment can be reproduced or predicted when adequate computational power is available. Techniques for the computation of pure solvents or dilute solutions, macromolecules, and solid-state systems are also advanced, but accuracy and size are serious limitations. Many efforts in chemical science have merged with biology or with materials science; such modeling requires accuracy for very large systems, and the ability to deal with complex, macromolecular, supramolecular, and multicomponent heterogeneous systems. The nanosystems revolution is very small from the perspective of size, but huge from the viewpoints of chemical engineering and chemistry because it will allow wholly different design and manufacture at the molecular level. This will comprise a new subdiscipline, molecular engineering.

Computational, communications, and data storage capabilities are increasing exponentially with time. Along with massive increases in computational power, better computational algorithms for large and multicomponent systems are needed if computations are to have a major role in design. Some of the major target opportunities for exploration in the near term include

- *Computational Methods:* The implementation of theoretical models within software packages has now become excellent for certain focused problems such as molecular electronic structure or simple Monte Carlo simulations. Very large challenges remain for extending these methods to multiscale modeling in space and time.
- *Education and Training:* The community has not made as much progress as might be desired in the training both of chemical professionals and professionals in other disciplines. Training techniques tend to focus on particular packages, lack integration, and be concentrated too narrowly within particular subdisciplines. Education in chemistry and chemical engineering has not yet utilized IT advances in anything approaching a reasonable way.
- *Databases:* The database resource in the chemical sciences is rich but fragmented. For example, the structural databases for small molecules and for biological systems are in an early stage of development and integration. Often the data have not been verified, and federated databases are not in widespread use.

- *Ease of Use:* The field has a very bad record of resuscitating "dead code"—code that has been modified so that some branches will never be executed. Too many of the software packages in current use, both commercial and freeware, are very poor from the viewpoints of reliability, user friendliness, portability, and availability. Maintenance and improvement of codes is not well handled by the field in general.
- *Problem Solving:* Particular areas such as drug design, molecular structure prediction, prediction of materials properties, energetics of particular reactions, and reaction modeling have shown major successes. These successes will provide the springboard for designing the Collaborative Modeling-Data Environments that constitute a major theme of this report.
- *Optimization:* Large-scale nonlinear optimization techniques for continuous and discrete variables are just beginning to make their way into every part of the chemical sciences, from the molecular level to the enterprise level. At the molecular level these techniques are being used to design molecular structures, while at the enterprise level they are being used to optimize the flow of materials in the supply chain of the chemical industry. Issues such as sensitivity analysis, parameter estimation, model selection, and generalized optimization algorithms are particularly important—but are not yet in common use.
- *Supply-Chain Modeling:* In this crucial area, security, privacy, and regulatory issues must be addressed. Supply-chain structures are characteristic of large-scale problems of importance both within chemical science and engineering and across the larger society. Just as the impact of such work will be major, so will the importance of assuring accuracy, security, privacy, and regulatory compliance. Often the data used in such simulations are proprietary and sensitive; thus, access to both the data and the models themselves must be controlled so that the effectiveness and societal acceptance of such modeling are not jeopardized.

CHALLENGES

In the future, modeling and data management will become unified. Research, development, education, training, and understanding will be done in a comprehensive multiscale problem-posing and problem-solving environment. Targeted design, as well as our understanding of simple systems, can profit by investigation within such a holistic research environment.

Although the exponential increases in capability and utilization cannot continue forever, the advances already attained mean that the capabilities available to the chemical sciences are not limited to commodity computing and the constraints of more, faster, and cheaper cycles. Major challenges exist in maximizing the ability of chemical scientists to employ the new methods and understandings of computer science:

- *Develop interdisciplinary cooperation with applied mathematicians and computer scientists to approach chemical problems creatively.* Particular targets would include use of informatics, data mining, advanced visualization and graphics, database customization and utility, nonlinear methods, and software engineering for optimization and portability of chemical codes.

- *Develop computational chemical sciences, as a focused subdiscipline of chemistry, chemical engineering, computer sciences, and applied mathematics.* Cross-disciplinary training grants, planning workshops, symposia, programming groups, and other forms of interaction sponsored by professional societies and funding agencies should materially help in bringing computer science and applied math expertise to solving advanced problems of chemical modeling and data utilization.

- *Better integrate the two disciplines of chemistry and chemical engineering.* These face nearly identical issues, concerns, advantages, and IT-based challenges. Integrated software, communications, database, and modeling capabilities can be used as a pathway for closer linking of these disciplines.

- *Generate graphical user interfaces.* Current advances in computer science permit adaptation of existing complex codes into a larger problem-solving environment. Tools for semiautomatic generation of new graphical user interfaces could facilitate calling or retrieving data from these extant programs. Eventually, one can envision an object-oriented approach that would include interfaces to a suite of components for integrating chemical science modeling tasks such as electronic structure calculation, smart Monte Carlo simulation, molecular dynamics optimization, continuum mechanics flow analysis, electrochemical process modeling, and visualization. The disciplinary expertise in chemistry and chemical engineering is a key element in making the interfaces broadly useful within the discipline.

- *Develop more general methods for manipulating large amounts of data.* Some of the data will be discrete, some stochastic, some Bayesian, and some Boolean (and some will be unreliable)—but methods for manipulating the data will be needed. This integrative capability is one of the great markers of human intelligence; computational methods for doing such integration are becoming more widespread and could transform the way that we treat the wonderful body of data that characterizes the chemical sciences.

- *Help the IT community to help the chemical world.* One issue is the need to modify the incentive and reward structure, so that optimization experts or data structure engineers will want to attack chemical problems. Another issue is the rapid exporting of IT offshore, while national security and economic considerations require that extensive IT expertise and employment also remain in the United States. A national award for IT professionals working in the chemical sciences would be useful as one component for addressing these issues. Another possibility would involve new postdoctoral awards for IT experts to work within the chemical sciences.

The preceding challenges involve bringing tools from information technology, in a much more effective fashion, to the solution of problems in the chemical sciences. There are also major problems within the chemical sciences themselves, problems that comprise specific challenges for research and technology development. The largest-scale challenges deal with the ability of chemistry and chemical engineering to address major issues in society. We should focus on major issues:

- *Provide stewardship of the land.* We need to develop new methods in green chemistry for manufacturing in the twenty-first century, and we need to accelerate environmental remediation of sites around the world that have been polluted over the previous century. Examples include
    o incorporating computational methods into sensors for real-time analysis and assimilating sensor-measurement-information data in simple yet technically correct formats for use by public policy decision makers; and
    o expanding our environmental modeling efforts—especially atmospheric and oceanic modeling—to account for the impact and fate of manufactured chemicals; to assess how changes in air and water chemistry affect our health and well-being; and to develop alternative, efficient, and clean fuel sources so that we need not rely on imported hydrocarbons as our major energy source.
- *Contribute to betterment of human health and physical welfare.* This challenge extends from fundamental research in understanding how living systems function as chemical entities to the discovery, design, and manufacture of new products—such as pharmaceuticals, nanostructured materials, drug delivery devices, and biocompatible materials with lifetimes exceeding patients' needs. This challenge is especially relevant to computing and information-based solutions, because we will need to develop smart devices that can detect and assess the early onset of disease states. Moreover, we need to convey that information in a reliable manner to health care professionals. We will also need to develop large-scale simulations that describe the adsorption, distribution, metabolism, and excretion of as-yet-unsynthesized drugs. We need to do this to account for naturally occurring phenomena such as bacterial evolution that make extant drugs ineffective, and we must be able to do it in quick response to potential homeland security disasters that affect our health and lives.
- *Ensure an informed citizenry through education.* Our democratic society depends on an informed citizenry, and a major challenge facing the chemical sciences community involves education. This includes reaching out to our colleagues in other disciplines who are beginning to study how things work at the molecular level. A need exists to integrate seamlessly chemistry with biology and materials science and to strengthen the connections between chemistry and chemical engineering. The challenges described above concerning environmental,

health and welfare issues are also challenges to the community of educators. The educational challenge includes public relations, where it will be necessary to address the negative connotations of the word "chemical." Educational efforts that use such information technology tools as websites and streaming video could help to deliver accurate information and to inform the level of discourse within the public domain.

- *Facilitate more thoughtful decision making.* We need informed policy and decision makers to guide our society in times of increasingly complex technological issues. Examples include international policies on health, disease, environment, natural resources, and intellectual property, to name a few. Issues that arise include mining databases for relevant information, developing and evaluating model scenarios, and assessing uncertainty and risk. The goal "every child a scientist" should be attainable, if by "scientist" we mean someone with an understanding of what science is.

- *Protect and secure the society.* Of all the scientific and engineering disciplines, the chemical sciences and technology have perhaps the strongest influence on the economy, the environment, and the functioning of society. The community of chemical engineers and scientists must retain responsibility both for maintaining the traditional intellectual and methodological advances that chemistry and chemical engineering have provided and for responsibly managing the footprint of chemistry. Issues of privacy and security are crucial here, as are scrupulous attention to responsible care and continued responsible development. Using information technology to achieve these goals will greatly enhance our ability to protect and secure both the principles of our discipline and the civilization in which that discipline is used and practiced.

Finding: There are major societal and civic problems that challenge the chemical community. These problems should be addressed by chemistry and chemical engineering, aided by IT advances.

These societal issues include providing stewardship of the land, contributing to the betterment of human health and physical welfare, ensuring an informed citizenry through education, facilitating more thoughtful and informed decision making, and protecting and securing the society.

The chemical sciences need to develop data-driven, natural teaching- and information-capture methods, preferably including some that do not require equation-based algorithms (what might be called natural learning). The community should develop means for using assembly-knowledge methods to produce weighted, customized data-search methodology (which would, effectively, correspond to designing a chemistry-engineering Google search engine). The community also should utilize such popular IT advances as web searching and 3-dimensional graphics developed for games. Finally, a verification or standardization scheme for data is needed.

Within the overall challenge themes just expressed, it is useful to focus on

> Sampling is a key bottleneck at present in obtaining accurate results in molecular modeling simulations. Obtaining convergence for a complex condensed-phase system is extremely challenging. This is the area in my opinion where prospects are most uncertain and where it is critical to support a lot of new ideas as opposed to just improved engineering of existing approaches. Some advances will come about from faster hardware, but algorithmic improvement should contribute even more if sufficient effort is applied.
>
> Richard Friesner (Appendix D)

several specific, targeted challenges that the chemistry and chemical engineering community will need to address:

- *Multiscale methodologies are crucial.* The community needs to incorporate advanced theoretical ideas to generate nearly rigorous techniques for extending accurate simulations to deal with phenomena over broad ranges of time and space. It also needs to attack such methodological problems as model-based experimental design, virtual measurement, quantum dynamics, integration with continuum environments, dispersion energetics, excited states, and response properties. All of these are part of the multiscale modeling capability that will be crucial if the chemical community is to utilize advanced IT capabilities in the most effective and productive fashion.

- *Optimization must go beyond studies of individual systems and deal with chemical processes.* The community must develop effective design, planning, and control models for chemical processes. It will be necessary to address the integration of these models across long time scales, as well as their accuracy and utility.

- *The community must develop enterprise-wide optimization models.* Using advanced planning and scheduling methods, the models must allow fast and flexible response of chemical manufacturing and distribution to fluctuations in market demands. The economic implications and societal importance of this effort could be enormous.

Finding: The nation's technological and economic progress can be advanced by addressing critical needs and opportunities within the chemical sciences through use of new and improved information technology tools. Bringing the power of IT advances to bear will greatly enhance both targeted design through multidisciplinary team efforts and decentralized curiosity-driven research of individual investigators. Both approaches are important, but they will depend upon IT resources in different ways. Among the needs to be addressed are:

- Take strategic advantage of exponential growth of IT resources, which will revolutionize education, research, and technology in chemical science and engineering.
- Develop a rational basis for dealing with complex systems.
- Embrace simulation, which is emerging as an equal partner with experiment and theory.
- Utilize web-based collaborative problem solving: Collaborative Modeling-Data Environments).
- Recognize the increasing importance of multiscale, multiphenomena computing.
- Maintain support for the fundamental aspects of theory that will remain essential for progress.
- Support an approach to algorithm development that is application driven.
- Facilitate the development of technologies for building reliable and secure multiscale simulations. Multiscale computational methodologies are crucial for extending accurate and useful simulations to larger sizes and longer times. Fundamental advances in the formalisms and methodologies, as well as in algorithmic software and visualization advances, will be needed to make effective multiscale modeling a reality.

Finding: To sustain advances in chemical science and technology, new approaches and IT infrastructures are needed for the development, support, and management of computer codes and databases.

Significant breakthroughs are needed to provide new means to deal with complex systems on a rational basis, to integrate simulations with theory and experiment, and to construct multi-scale simulations of entire systems.

It will be necessary to

- develop methods for semi-automatic generation of user interfaces for codes and design modules in the chemical sciences—with the eventual goal of a semiautomated object-oriented modeling megaprogram(s) containing modules for specific capabilities;
- develop reliable error estimators for computational results;
- develop enhanced methodology for data mining, data management, and data-rich environments, because databased understanding and insights are key enablers of technical progress; and
- develop improved methods for database management, including assurance of data quality.

4

Interfaces: Cooperation and Collaboration Across Disciplines

The research and technology enterprise of chemistry and chemical engineering involves a dazzling multitude of intellectual opportunities and technological applications. The community includes academia, industry, and government and independent laboratories, and it embraces the disciplines of chemistry, chemical engineering, and many others that interact in complex ways. Working at the interfaces between disciplinary domains offers reinforcing synergies and beneficial blurring of boundaries, but may also be accompanied by barriers that hinder the effectiveness of information flow. Information technology (IT) assets are rapidly accepted in the chemical sciences, and they are used today mainly to increase the speed of what we already know how to do. The strategic use of computing resources requires deeper integration of chemical engineering and chemistry with IT, a process that is in its early stages.

Finding: Computation and information technology provide a key enabling force for lowering barriers among the disciplines that comprise the chemical enterprise and closely related fields.
Identification of mutual interests among disciplines and removal of the barriers to successful communication among constituencies are essential for increasing the overall effectiveness of the system. The processes of identification and removal are still in their infancy.

Cooperation and collaboration across disciplines expand the intellectual freedom that people enjoy—the freedom to access the tools and capabilities to do one's work. The interface between information technology and the activities of chemists, chemical engineers, and allied disciplines is still in its infancy. The opportunity to build that architecture, which will revolutionize the future of the chemical enterprise, is exciting. It is essential that advanced information systems draw on and reinforce the intuitions and creative intellectual patterns that have been so successful in research, discovery, and technology innovation in the chemical enterprise. Moreover, it is critically important to ensure the security of transactions that involve sharing of tools, data, and resources in a manner that provides responsibility for privacy, intellectual property, economic growth, and the public trust.

Intuition based on physical understanding of the system is an essential component for integrating computation and information technologies with the chemical sciences. The essential resource driving the interface of IT and the chemical sciences is human ingenuity, and the supply of this resource is unbounded. Indeed, throughout this report, the importance of "people issues" emerges. Cooperation and collaboration depend on the attitudes of the individuals who participate. An important first step toward cooperation and collaboration is training that nurtures a creative, problem-solving approach. The cross-disciplinary interface between IT and the chemical sciences should be capable of rejecting preconceptions about the way things "ought to be," while also providing stable tools that support advances in research and technology. People, not software or data repositories, will recognize the revolutionary breakthrough and how to apply it to the right problem. People will recognize how a suite of interlocking technologies can be combined with revolutionary impact.

Finally, it is important to have realistic expectations for success. This is a difficult challenge because what one measures and rewards is what one gets. There are many ways to impact applications and technology with any level of sophistication in a simulation. Some of the important ways lend themselves only to intangible measures, but oftentimes these may be the most important. Again quoting Einstein, "Not everything that counts can be counted, and not everything that can be counted counts."

Ellen Stechel (Appendix D)

Research discoveries and innovative technological advances invariably accompany breakthroughs in computation and information technology. The advances to date in these areas make possible an even faster pace of discovery and innovation. Virtually every one of the discovery and application opportunities mentioned in the previous chapter depends on sustained advances in information technology.

The opportunities at the chemical science–IT interface are enormous. Advances in each of the IT areas discussed here will benefit virtually all of the discovery and application areas described. Profound advances will result when developments in the IT area are brought to bear on discovery and applications areas. It is critically important to learn how to build teams that bring the component disciplinary skills to bear. It is equally important to maintain depth and disciplinary skills in the fundamentals. Clearly the impact gained by solving a key IT challenge in one particular area can be significantly multiplied if it is carried out in an informed manner that benefits related areas.

The process of bringing IT developments to bear on discovery areas can serve also to guide new research within the IT discipline. The close interaction of IT experts with science and engineering application experts requires a management approach that is in part maintaining the value systems that cross multiple disciplines, and in part "enlightened self-interest." The major challenge is to set clear investment priorities, and to maximize the benefits across many applications.

The computation-information infrastructure within chemical sciences is variegated and effective. New directions, based on both IT advances and the growth of interdisciplinary understanding, offer striking possibilities for research development, learning, and education. One specific new endeavor, the establishment of integrated, multicomponent Grid-based Collaborative Modeling-Data Environments (CMDEs), is a promising opportunity for federal investment.

OVERARCHING THEMES

Four key themes emerge repeatedly at the heart of discussions on integrating information technology with the chemical sciences. These themes provide a framework for relating research areas with the enabling IT technologies. They are (1) *targeted design and open-ended discovery*—two routes to solving complex problems; (2) *flow of information between people within and among disciplines*—people issues and barriers to solving complex problems; (3) *multiscale simulation*—one of the key IT-centric technical approaches to solving complex problems; and (4) *collaborative environments*—integrating IT methodology and tools for doing the work. Appropriate educational initiatives linked with these themes are a major need.

Targeted Design and Open-Ended Discovery

Targeted Design. Targeted design involves working backward from the desired function or specifications of a molecule or product and determining the underlying structure and the process conditions by which it can be fabricated. The general features of the approach have been used successfully in cases for which the need was exceptional and the target was clear: for example, synthetic rubber, AIDS treatment, and replacements for chlorofluorocarbon refrigerants. Extraordinary recent advances in computer speed, computational chemistry, process simulation and control, molecular biology, materials, visualization, magnetic storage, bioinfomatics, and IT infrastructure make possible a wholly new level of implementation. By building on past investments, targeted design could pay enormous dividends by increasing the efficiency of the discovery effort and speeding up the innovation process. The process can also provide fundamental researchers with intuitive insights that can lead to more discoveries, However, a number of questions will need to be addressed as this approach is used more widely:

- *Structure-Function Relationships:* Given the molecular structure, how can its properties be estimated? Given the properties, how can they be related to the desired structure?
- *Combinatorial Search Algorithms:* How do we evaluate many alternatives to find the ones that meet specifications?
- *Target Identification:* What are the criteria for specifying the target?
- *Roles of Participants:* How does the technical problem translate into what the participants must do to address the issue successfully?
- *Collaborative Problem Solving:* What tools are needed to connect individuals from different disciplines, and how can these be made to reflect the way people work together?
- *Trust and Confidence:* Appropriate security is essential if collaborations involve proprietary information or methodology. How can the key elements of trust and confidence be assured?

Open-ended Research and Discovery. Open-ended research and discovery has a long tradition and spectacular record of success in the chemical sciences. Curiosity-based discovery has led to research discoveries (such as oxygen, nuclear magnetic resonance, DNA structure, Teflon, and dimensionally stable anodes) that have truly changed the world. The broad-based research investment of the past half century in curiosity-driven research is still paying great dividends in technological developments in such areas as nanoscience and nanotechnology, microelectromechanical systems (MEMS), environmental chemistry, catalysis, and the chemical-biological interface. Such an approach requires:

- access to information in a way that informs intuition;
- the flexibility for the individual investigator to modify the approach (an avenue of research in which open-source software can be of tremendous value); and
- data assimilation and visualization (where better and more efficient algorithms for mining data are needed, and where the chemical sciences must learn from the information sciences).

Targeted design builds on the foundation of previous curiosity-driven discoveries, but its progress also drives the need for new fundamental understanding. To view the technological developments that grow from open-ended research as a series of coincidences is to miss the point of the past investment strategy of many federal funding agencies. Both are essential and both benefit from advances in information technology. To take full advantage of the discoveries in such areas as chemical biology, nanoscience, environmental chemistry, and catalysis requires both approaches. It is important to distinguish between them:

- Targeted design does not work if we do not know the target; it is not open ended.
- Curiosity-driven research rarely hits a technology target on its own without guidance; the rate of innovation can be very slow.
- The common ground between these approaches can be expanded and enriched by IT advances.

The targeted design approach is beginning to be used in the IT community. Examples are available that are built on several decades of real success in computer science and in pure and applied mathematics:

- *Very advanced special-purpose computers*, such as the Earth Simulator in Japan; its raw speed is high, as is effectiveness of delivered performance. The machine takes advantage of memory-usage patterns in science and engineering applications by way of fast data paths between the central processing unit and main memory (cf. T. Dunning, Appendix D).
- *Algorithmic advances:* Today's supercomputers use commercially oriented commodity microprocessors and memory that are inexpensive but have slow communications between microprocessor and memory. They try to minimize slow access to main memory by placing a fast-cache memory between processor and main memory. This works well if algorithms can make effective use of cache, but many chemical science and engineering algorithms do not. Cache-friendly algorithms are needed to take full advantage of the new generation of supercomputers.
- *Algorithm development based on understanding* of the physical system as well as the computer architecture.
- *Development of "model" sciences* to obtain consistent results without unnecessary computation: This approach is illustrated by the Gaussian-2 (G2) model chemistry developed by Pople's group.[1]

The use of targeted design to bring IT resources to bear on problems in chemical science and technology is likely to become more widespread. That approach would provide progress for virtually every point identified earlier in this chapter. Many of the applications require capacity computing—linked with robust, reli-

---

[1] Curtiss, L. A.; Raghavachari, K.; Pople, J. A. *J. Chem. Phys.* **1995**, *103*, 4192.

Appropriate Computational Levels

I have a comment about your discussion of the convergence with respect to basis set in the computation or calculation of molecular energies. Probably the most notable feature of that is it became very expensive as you moved to the larger and larger basis sets but the last order of magnitude in computing that you in fact spent gave you almost no information, a result differing very little from the previous numbers, and in fact it seems to me that one is in some danger of exaggerating the need for very large computation to achieve chemical accuracy.

Many of the results that we have been able to find, we have found can be achieved by the judicious use of small empirical corrections or simple extrapolation schemes. I think that was already evident from your presentation. So I feel there is a slight danger in saying that you can only achieve this high a level of accuracy by using massive resources.

John A. Pople, Northwestern University
(comments following presentation by Thom Dunning, Appendix D)

able, and accessible problem-solving tools that gain leverage from rapid advances in network bandwidth, storage, and algorithm efficiency. The need for increased capability high-performance computing will also continue to be critically important for the long term, as larger, more complicated, and more accurate simulations become needed. In most cases, increases in performance are a result of adroit algorithms based on physical understanding of the application in addition to raw processor speed.

Flow of Information Between People Within and Among Disciplines

Advances in communication and information technology have helped make it possible to address complex systems with unprecedented success. Such systems exhibit nonlinear relationships, large experimental domains, and multiple interactions between many components. Moreover, for most complex systems, the underlying model is generally unknown although some components may be well-characterized. It is therefore necessary to employ various kinds of knowledge, theory, insight, experimental data, numerical codes, and other tools. In general, the flow of information passes among communities that may be organized by skill set (experimentalist, theorist, modeler, programmer, etc.); institution (academia, industry); discipline (chemistry, chemical engineering, computer science); or other such broad groupings.

INTERFACES: COOPERATION AND COLLABORATION ACROSS DISCIPLINES 35

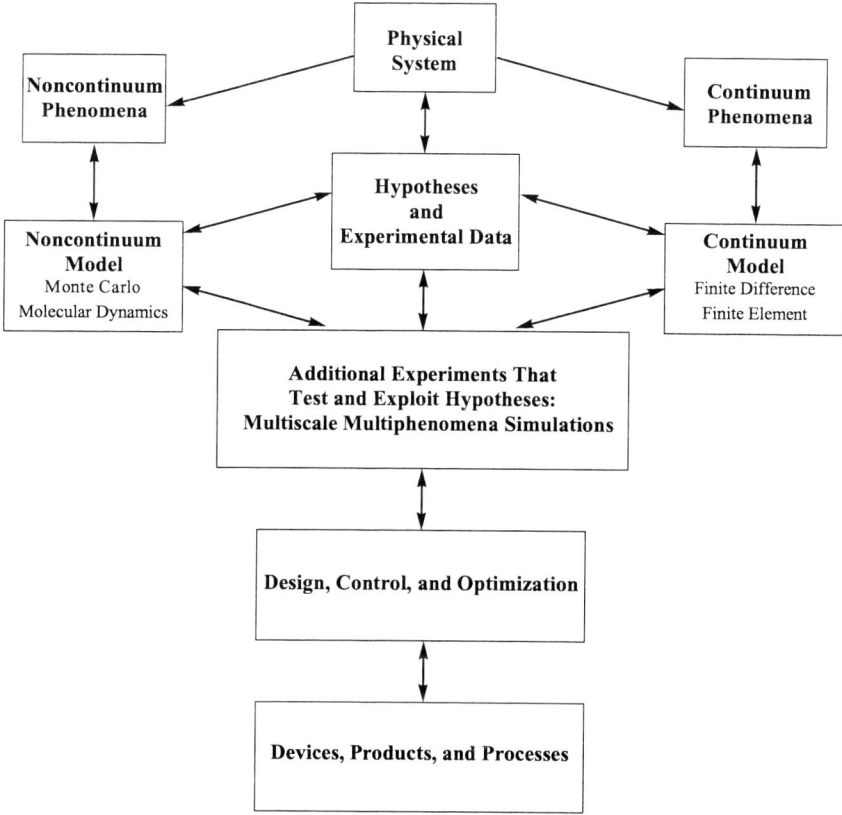

FIGURE 4-1 Flow of information between science and technology in multiscale, multiphenomena investigations.

Figure 4-1 describes some aspects of information flow in the analysis of complicated problems. The boxes in the figure represent specialized methods and tools that contribute to the investigation. These include, for example, experimental techniques, computer codes, numerical methods, mathematical models of well-characterized fundamental scientific systems, and numerical simulations of complex components of technological systems. Several general observations can be made:

• The tools used in one box often are not easily used by specialists working in another box.
• Robust, reliable methods of remote access are now beginning to emerge: shared equipment (e.g., NMR), data (e.g., bioinfomatics), and codes (e.g., open source). Such beginnings strengthen the flow of information between some of the boxes.

- Between other boxes, however, barriers still hinder the effectiveness of the information flow as noted in the next paragraphs.

For the purpose of discussion, consider a specific starting point: *Ensuring product quality at the molecular level requires a new generation of science and engineering tools.* This statement applies to many of the areas discussed in this report, from nanoscience to biotechnology to sensors and microelectronic devices. The following material considers how this statement translates into needs and barriers between two groups: the technical experts (who create the methods and tools inside the boxes) and the users (who use the methods and tools to solve problems). First consider the roles of different kinds of technical experts. It is abundantly clear that the following characterizations are incomplete and overly simplified, but they nevertheless serve to illustrate some of the salient needs of, and barriers between, different communities of technical experts:

- *Experimentalists* want to produce reliable data. To interpret and validate the data with sophisticated computational codes, they need modeling expertise. To test the data and to make it accessible to others, they need computer science expertise in database structures.
- *Modelers* want to create simulation codes that compile and run efficiently. From experimentalists, they need up-to-date mechanisms and data. From computer scientists and engineers, they need to understand the language, advancing programming methods, protocols, data management schemes, and computing architectures, to name a few.
- *Computer scientists and engineers* want to push the envelope of IT capabilities, and create structures for efficient workflow and resource utilization. They need metadata from experimentalists in order to characterize the data and facilitate use of the data by others. They need an understanding of the physical basis of modeling goals, which are invariably the key to improving efficiencies.

Figure 4-2 provides a schematic representation in which the three kinds of technical expertise are at the corners of a triangle, and the needs that they have from each other are summarized along the sides. Barriers are shown that block information flow between the experts.

Once the barriers are removed, each group will want to change the way it does things in order to take full and strategic advantage of the others. In keeping with a focus on the statement that *ensuring product quality at the molecular level requires a new generation of science and engineering tools,* the following examples serve as illustrations:

- *Experimentalists* will produce data at small scales on well-characterized systems to validate numerical simulations, and also provide measurements at

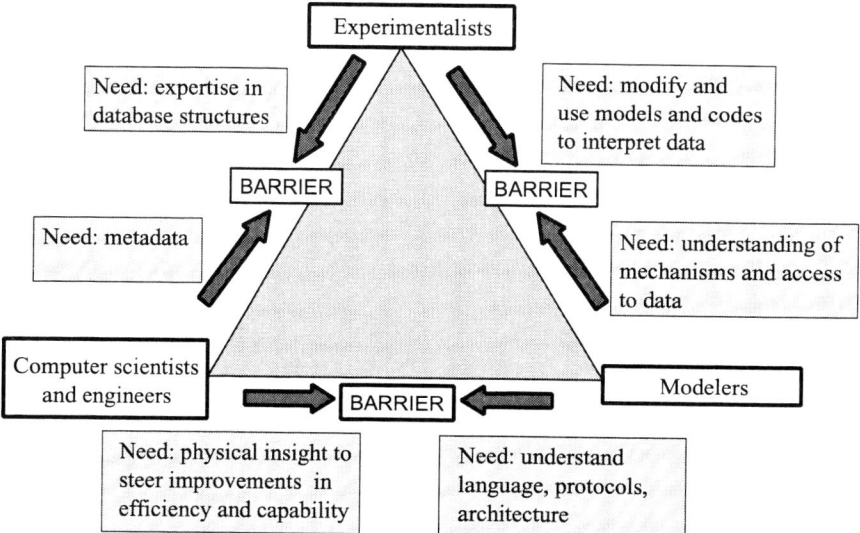

FIGURE 4-2 Needs and barriers for interactions among technical experts.

multiple scales on entire complex systems to characterize sources of uncertainty. They will document error bars on datasets including images so that others can use them with confidence. They will be able to suggest simulations to test or resolve multiple reasonable hypotheses of mechanisms and use up-to-date data from others in those simulations.

• *Modelers* will produce portable object-oriented codes so others (including nonexperts) can modify the original hypotheses and assumptions. They will combine theories at different scales with metadata schema to link continuum and noncontinuum codes so that multiscale simulations can be done routinely. They will be able to draw on different experimental methods and data for improved simulation including parameter sensitivity and estimation, hypothesis selection, lifetime probability, and risk assessment. They will tune code performance based on understanding of computer architecture and algorithm strategies.

• *Computer scientists and engineers* will develop robust, reliable, user-friendly Collaborative Modeling-Data Environments for linking codes, computers, research data, field-sensors, and other inputs for use by scientists and engineers. They will provide semiautomatic generation of graphical user interfaces so that older codes will continue to be useful for design and synthesis as new computing resources become available. In addition, improved security, authentication, authorization, and other measures will be developed that protect privacy and intellectual property associated with commercial applications. The use of web services will be crucial for the utility of this work.

In order to reduce barriers several steps must be addressed:

- Promote and ensure decentralized autonomy of creative technical experts in the chemical sciences, while facilitating a multidisciplinary team approach to targeted design.
- Continue the integration of IT with the chemical sciences, which have been quick to accept IT and use it throughout the discipline.
- Increase multidisciplinary training and provide financial mechanisms to support IT.
- Develop modular, portable, and extensible programming methods, including object-oriented component architectures for portable modeling (i.e., software that can be reused in multiple applications).
- Expand use of open-source software for community codes where it is appropriate for many individuals to participate in code development.
- Strengthen the scientific method of testing hypotheses by experiment by lowering barriers to flow of information among modelers, experimentalists, and computer scientists.
- Optimize the research infrastructure for multidisciplinary team approaches to problems.
- Develop a permanent, ongoing interaction between the chemical sciences and IT communities to guide investments in infrastructure for the chemical sciences.

Next consider the roles of users who use the methods and tools developed by technical experts to solve problems.

- Users may include other technical experts, industrial technicians and trainees, students, and learners.
- Most users will want to input their own variables—which may be confidential—and will want return to their work leaving no fingerprints for others to view.
- Such users may have limited interest in technical details, but will need a clear understanding of assumptions and limitations on use as well as robust, reliable, user-friendly access.
- Users will require a feedback mechanism so they can direct the experts to areas where improvements are needed.

Finding: Addressing critical challenges at the interfaces with other scientific and engineering disciplines will enable chemistry and chemical engineering to contribute even more effectively to the nation's technological and economic progress.

The most important challenge involves people. Advances in IT that facilitate self-organization of problem-solving groups with common interests across disciplinary boundaries will impact strongly both understanding-based and application-driven projects. The essential resource driving the interface of IT and the chemical sciences is human ingenuity.

Multiscale Simulation

After a century of success in understanding the fundamental building blocks and processes that underlie our material world, we now face another, even greater, challenge: assembling information in a way that makes it possible to predict the behavior of complex, real-world systems. We want to predict properties of physical, environmental, and biological systems over extended length and time scales; understand how component processes interact; fill gaps in experimental data with reliable computational predictions of molecular structure and thermodynamic properties; and develop reliable methods for engineering process design and quality control. Here is a scientific frontier for the twenty-first century. With virtually any of the examples emphasized throughout this report, multiscale simulation will play a critical role.

Figure 4-3 shows the broad range of time and length scales—some 15 orders of magnitude or more—that are encountered for one particular field of application involving electrochemical processing for chip manufacture. The figure is divided into three vertical bands in which different kinds of numerical simulation tools are used: noncontinuum, continuum, and manufacturing scale. Each of the topics indicated in the figure is placed in the approximate position to which it corresponds. For some of the topics, the space and/or time axis may be more or less approximate and appropriate. There is today a sophisticated understanding of each of these topics. Multiscale simulation involves linking the pieces in order to understand interactions within an entire system.

Other topics such as drug development and manufacture, environmental remediation, or coatings technology show similar multiscale aspects. A few examples serve to illustrate the very broad range of activities for which multiscale simulations are important:

- The internal combustion engine involves the engine itself, as well as the fluid dynamics of the fuel-air mixture as it flows through the combustion region and the chemical process of ignition and burning of the fuel, which can involve hundreds of chemical species and thousands of reactions, as well as particulate matter dynamics and the ultimate fate of combustion products through atmospheric chemistry.
- Creating a virtual simulation of a periodic living cell involves a substantial challenge. The organelles in such a cell include about 4 million ribosomes. There are about 5 billion proteins drawn from 5000-10,000 different species, a meter of DNA with several billion bases, 60 million transfer RNAs, and vast

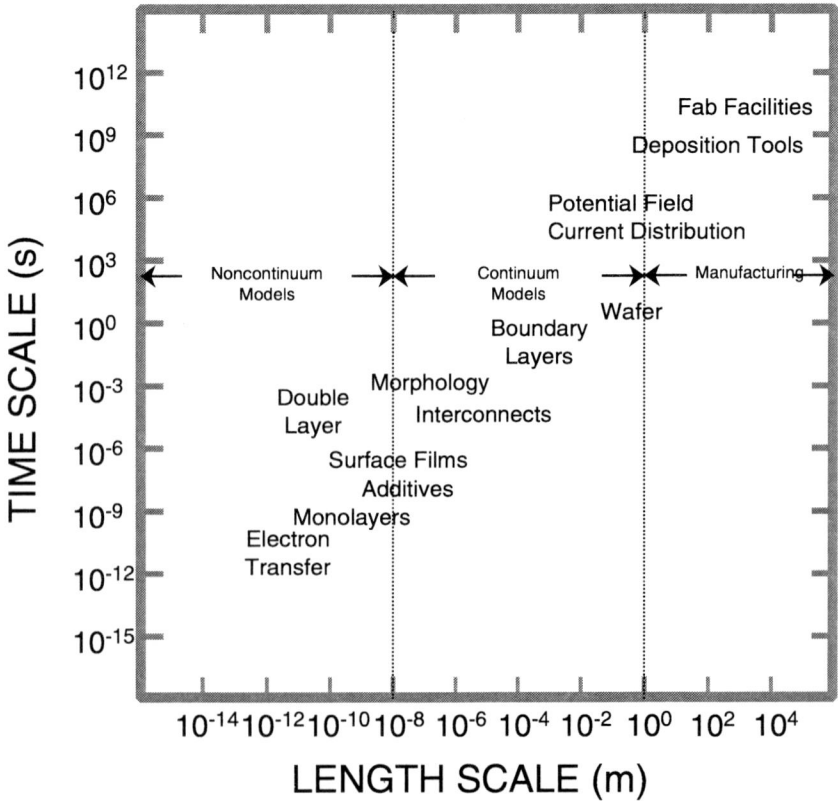

FIGURE 4-3 Schematic of the time and length scales encountered in multiscale simulations involving electrochemical processing for chip manufacture.

numbers of chemical pathways that are tightly coupled through various feedback loops in which the proteins are turning the genes on and off in the regulation network. All of the relevant detail would have to be captured.

• The design and degradation of materials involves multiple scales—from atoms to processes, as well as stress and environmental degradation from corrosion, wear, fatigue, crack propagation, and failure.

Multiscale simulation involves the use of distinct methods appropriate for different length and time scales that are applied simultaneously to achieve a comprehensive description of a system. Figure 4-4 illustrates some of the computational methods that have been developed over many decades in order to deal with phenomena at different time and length scales to compute properties and model phenomena. These include *quantum mechanics* for accurate calculation of small

Microscale Design Issues

The major development that is driving change, at least in my world, is the revolution at the micro scale. Many people are working in this area now, and many think it will be as big as the computer revolution was. Of particular importance, the behavior of fluid flow near the walls and the boundaries becomes critical in such small devices, many of which are built for biological applications. We have large molecules moving through small spaces, which amounts to moving discrete molecules through devices. The models will often be discrete or stochastic, rather than continuous and deterministic—a fundamental change in the kind of mathematics and the kind of software that must be developed to handle those problems.

For the engineering side, interaction with the macro scale world is always going to be important, and it will drive the multiscale issues. Important phenomena occurring at the micro scale determine the behavior of devices, but at the same time, we have to find a way to interact with those devices in a meaningful way. All that must happen in a simulation.

Linda Petzold (Appendix D)

---

and fast phenomena, *statistical mechanics* and semiempirical modeling for mechanistic understanding, the *mesoscopic scale* where new methods are now beginning to emerge for multiscale modeling, *continuum mechanics* for macroscopic reaction and transport modeling, *process simulation* for characterizing and optimizing entire process units and their interconnection, and *supply chain modeling* that integrates over multiple plants, customers, and global logistics. In many cases, modeling and simulation techniques that were originally developed in other communities, such as computational physics and materials science, are now being used in the chemical sciences, and vice versa. Specific examples can be found in Appendix D in presentations by de Pablo, Dunning, Maroudas, and Petzold.

Ultimately what one would like to do from an engineering viewpoint would be to use all of these methodologies to explore vast regions of parameter space, identify key phenomena, promote these phenomena that lead to good behavior, avoid the phenomena that lead to failure, and so on. The range of scales may extend from quantum chemistry to atomistic and molecular modeling to mesoscopic to continuum macroscopic mechanics to large-scale integration of processes. Progress over the past decade at each of these scales suggests that it may now be possible to combine methods from different scales to achieve a comprehensive description of a system. Such techniques will address important issues

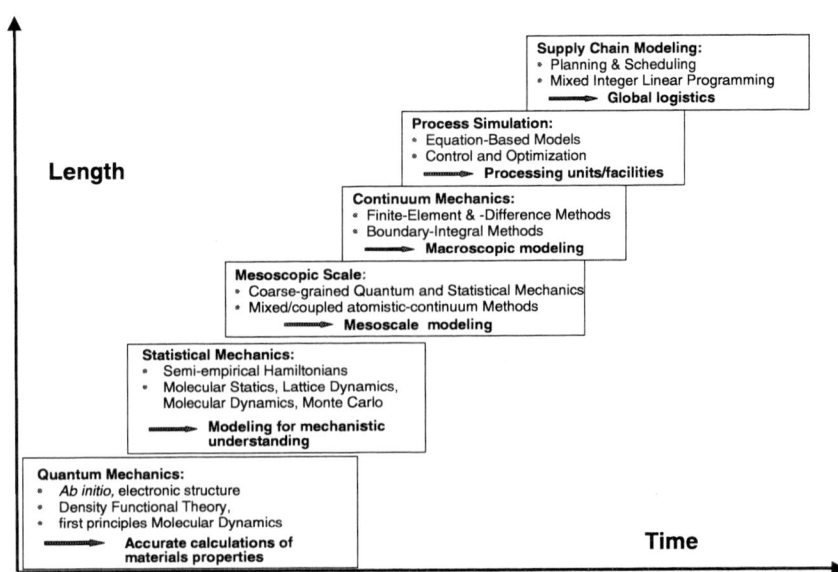

FIGURE 4-4 Modeling elements and core capabilities.

that include high-throughput, combinatorial modeling and high-dimensional design space. The simulation of systems that involve such broad ranges of length and time scales naturally requires a multidisciplinary approach.

The current generation of multiscale modeling activity represents the beginnings of a new approach for assembling information from many sources to describe the behavior of complicated systems. The current methodology will need steady improvement over the next decade. For any multiscale model, what is missing in the early stages of development is less important. What is important is what is *not missing*. Multiscale modeling is a kind of targeted design activity, and it requires physical intuition to decide what does not matter. These examples point to several critical areas where advances are needed to provide improved multiscale capabilities:

• *Computing Power:* The underlying progress of the past decade has resulted in part from the increasing power of computers and from important theoretical and algorithmic developments. Bringing these together for commodity computing places significant demands on the information infrastructure.

• *Computers:* Multiple architectures will be needed for different types of calculations. Major gains in the chemical sciences will require accuracy and an understanding of the error bounds on the calculations. High-end computing will increasingly serve to validate simpler approaches.

- *Formalisms:* Actual multiscale models are nearly always based on approximate separations of length and time scales (e.g., Langevin equations for coarse graining in time, or quantum mechanics–molecular mechanics algorithms for large systems). Formal approaches are needed for understanding and improving such approximations.
- *Software:* Investments in software are needed, especially in the area of automated programs for software development and maintenance. For multiscale simulations, component-based methods will be needed to develop simulations of entire systems. The funding portfolio for algorithm development should include application-driven projects that can establish the scientific and technological foundations necessary for a software component industry.
- *Interoperability:* Protocols and data structures are needed to achieve interoperability of applications that run concurrently.
- *Web Services:* Programs and associated data that can be accessed by other computers.
- Computational steering: The modeler should be able to interact with a simulation in real time to steer it towards a particular goal. This will require both algorithmic and hardware advances.
- *Uncertainty and Reliability:* Methods are needed for assessing sources of error in components and their aggregation into multiscale systems, for example, systems that combine computational results with experimental data and images.
- *Accessibility:* Object-oriented or other portable component architecture is needed for models that can be modified and reused by others.
- *Standardization:* Experts need certification standards for producing robust, reliable components that have a clear description of approximations and limits so that nonexperts can use them with confidence.
- *People:* A robust supply of workers with multidisciplinary training will be needed in response to increasing demand.

Many of the processes of interest in chemistry and chemical engineering occur on much longer time scales (e.g., minutes or hours); it is unlikely that the several orders of magnitude that now separate our needs from what is possible with atomistic-level methods will be bridged by the availability of faster computers. It is therefore necessary to develop theoretical and computational methods to establish a systematic connection between atomistic and macroscopic time scales. These techniques are often referred to as multiscale methods or coarse-graining methods.

Juan de Pablo (Appendix D)

Collaborative Environments

The most difficult challenges in advancing human welfare through knowledge of the chemical sciences involve a level of complexity that can only now begin to be addressed through information technology. Such efforts are inherently multidisciplinary and involve working in teams to assemble information and solve problems. New methods for collaborative discovery and problem solving in chemical science and technology are needed so that individual researchers can participate in distributed collective action and communicate effectively beyond their discipline. These methods should be a significant component of education in the chemical sciences.

Prototype collaborative environments are emerging, with examples such as simple web-based browsers that provide access to applications. The environments extend to more sophisticated demonstrations of Grid-based services that utilize advanced middleware to couple, manage, and access simulation codes, experimental data, and advanced tools, including remote computers. Current early prototype examples include the DOE Science Grid, EuroGrid (European Union), UNICORE (German Federal Ministry for Education and Research), Information Power Grid (NASA), TeraGrid Alliance Portal (NSF/NCSA), PUNCH (Purdue), and many other projects, applications, and libraries including some commercial activities. Such environments promise an integrating structure for composing substantial applications that execute on the Grid. They offer great promise for sharing codes, data, and expertise, as well as for linking the pieces and keeping them up to date. Many of the prototype environments utilize complex Grid-based protocols that are still in early stages of development or consist of demonstrations that are not easily reused or modified. If these prototypes are to fulfill their promise for effective problem solving, pioneering chemical scientists and engineers must be brought into close working relationship with computer scientists and engineers to develop and tune the architecture. Infrastructure expenditures in information technology will be investments only if they can eventually be used for solving problems faced by society.

To support growing demand for and dependence on information infrastructure for applications in the chemical science and technology, advances are needed in many different dimensions:

- *Software:* For collaborations, there will be increasing needs for user-friendly and robust software that provides assured standards for quality, reliability, access, interoperability, and sustained maintenance. Standards are needed to facilitate collaborative interactions between applications personnel who are not experts in computer science. Use of software to upgrade *legacy codes* (or *dusty decks*)—codes that have been around a while (10-15 years), usually written by others who are no longer available, run on current-generation computers and systems for which it was not optimized, and therefore difficult to modify and debug—will be increasingly important for sustaining the effective lifetime of collaborative tools so as to recover a larger portion of the initial investment.

- *Computers:* Because the range of applications and collaborative dimensions is large, access to a wide range of computing platforms is essential for the chemical sciences. In the past, federal investments in high-end computing and networking have provided a successful path for pioneering and eventually developing capabilities for a very broad-based user community. As the pace of change quickens in the future, it will become increasingly important to maintain even closer ties between pioneering efforts and their rapid dispersal in broad user communities.
- *Simulation Codes:* Today, the portfolio of commercial, open-source, and other specialty simulation codes constitutes a rich but unconnected set of tools. While some commercial and open-source codes are user-friendly, many specialty codes are idiosyncratic and excessively fragile. Advanced techniques, such as object-oriented programming, are needed to facilitate code reuse by individuals other than the creators. Working in collaborative groups, experimental scientists will want to modify codes (to test alternative hypotheses, assumptions, parameters, etc.) without the need to recompile. In addition, technologists will want to explore operating conditions and design variables, as well as scientific hypotheses.
- *Data and Knowledge Management:* Hypotheses often continue to influence decisions—even after being proven wrong. Legacy simulations too often operate on the basis of disproved or speculative hypotheses rather than on recent definitive data. The chemical enterprise has created extensive data crypts of flat files (such as Chemical Abstracts). It now needs to develop data-rich environments (such as fully normalized databases) to allow new cross-disciplinary research and understanding. Within certain industries, such capabilities already exist, at least in early form. An example is offered by the pharmaceutical industry, where informatics hubs integrate biological, chemical, and clinical data in a single available environment. The science and technology enterprise needs a way to manage data such that it is relatively easy to determine what the community does and does not know, as well as to identify the assumptions underlying current knowledge.
- *Uncertainty and Risk:* In collaborative efforts it is important to assess sources of error in both experimental and numerical data in order to establish confidence in the conclusions reached. Managing the uncertainty with which data are known will reduce the risk of reaching unwarranted conclusions in complicated systems. Improved, easy-to-use risk assessment tools will assist in identifying the weakest link in collaborative work and help steer efforts toward developing improvements where they are most needed.
- *Multiscale Simulation:* Many collaborative efforts involve multiscale simulations, and the comments about uncertainty and risk apply here as well.
- *Quality:* Models and data inform intuition but rarely persuade individuals that they are wrong. Careful design of Collaborative Modeling-Data Environments—to facilitate appropriate access to codes, data, visualization tools, com-

puters, and other information resources (including optimal use of web services)—will enable participants to test each other's contributions.

- *Security:* Industrial users in the chemical sciences require security in various ways, such as avoiding unwanted fingerprints on database searches, preventing inappropriate access through firewalls, and providing confidence that codes work and that data are accurate. Collaborative environments must facilitate open inquiry about fundamental understanding, as well as protect intellectual property and privacy while reducing risk and liability. Economic models differ among various user and creator communities for software and middleware—from open-source, to large commercial codes, legacy codes behind firewalls, and fleet-footed and accurate specialty codes that may, however, be excessively fragile. The security requirements of the industrial chemical technologies thus represent both a challenge and an opportunity for developing or marketing software and codes. In multidisciplinary collaborations, values including dogmatic viewpoints established in one community often do not map onto other disciplines and cultures.
- *People:* Collaborations are inherently multidisciplinary, and the point has been made throughout this report that the pool of trained individuals is insufficient to support the growing opportunities.

Collaborations bring many points of view to problem solving and offer the promise of identifying barriers rapidly and finding new unanticipated solutions quickly. Anticipated benefits include the following:

- bringing in simulation at the beginning of complex experiments to help generate hypotheses, eliminate dead ends, and avoid repeated failures from a trial-and-error approach;
- arriving at approximate answers quickly with a realistic quantification of uncertainty so that one has immediate impact, deferring highly accurate answers to a later time when they are needed; bringing in data at the beginning of simulation efforts allows simple calculations to help frame questions correctly and assure that subsequent detailed experiments are unambiguous; and
- attaining new capability for dealing with the realistic, not overly idealized, problems at various levels of sophistication; some problems lend themselves only to intangible measures, but often these may be the most important approaches.

Federal support for one or more Collaborative Modeling-Data Environments could impact tremendously the value of IT for the chemical community. These would be ideal structures for advancing learning, research, insight, and development on major issues confronting both the chemical community and the larger society. The CMDEs can offer advantages of scale and integration for solving complicated problems from reaction modeling to corrosion. These are also uniquely promising for computational tools and methods development, for developing integrated Grid-based modeling capabilities, for building appropriate metaprograms for general use, and for providing powerful computational/understanding structures for chemical professionals, students, and the larger community.

EDUCATION AND TRAINING

For the chemical sciences to be able to train and enable people in optimal fashion, we must understand and employ information and communications in a far more sophisticated and integrated fashion than we have done to date. This use of information and communications as an enabler of people—as a developer of human resources—is a great challenge to the chemical sciences in the area of information and communications.

In broad terms, information and communications methods should permit the chemical scientist to create, devise, think, test, and develop in entirely new ways. The interactions and cross-fertilization provided by these methods will lead to new materials and products obtained through molecular science and engineering. They will lead to understanding and control of such complex issues as combustion chemistry and clean air, aqueous chemistry and the water resources of the world, protein chemistry and drug design, and photovoltaics and hydrogen fuel cells for clean energy and reduced petroleum dependence. They can help to design and build structures from microelectronic devices to greener chemical plants and from artificial cells and organs to a sustainable environment.

Finding: The capability to explore in the virtual world will enable society to become better educated and informed about the chemical sciences.
Conveying the intellectual depth, centrality, societal benefits, and creative challenges of molecular systems will be greatly facilitated by the use of modeling, visualization, data manipulation, and real-time responses. All of these new capabilities will provide unparalleled environments for learning, understanding, and creating new knowledge.

Finding: The growing dependence of the chemical enterprise on use of information technology requires that chemical professionals have extensive education and training in modern IT methods.
This training should include data structures, software design, and graphics. Because data and its use comprise such important aspects of chemistry and chemical engineering, and because appropriate use of IT resources can empower unprecedented advances in the chemical arena, it is crucial that the appropriate training, at all levels, be a part of chemical education.

The Collaborative Modeling-Data Environments could both provide these capabilities and educate people from 8 to 88. The proposed CMDEs would allow unique learning by permitting students to use different parts of the metaprogram's capabilities as their understandings and abilities advance. This could give students specific problem-solving skills and knowledge ("I can use that utility") that are portable, valuable, and shared. All of these new capabilities would provide unparalleled environments for learning, understanding, and creating.

5

Infrastructure: Capabilities and Goals

Remarkable advances in information technologies—computer speed, algorithm power, data storage, and network bandwidth—have led to a new era of capabilities that range from computational models of molecular processes to remote use of one-of-a-kind instruments, shared data repositories, and distributed collaborations. At the current pace of change, an order-of-magnitude increase in computing and communications capability will occur every five years. Advances in information technology (IT) allow us to carry out tasks better and faster; in addition, these sustained rapid advances create revolutionary opportunities. We are still at the early stages of taking strategic advantage of the full potential offered by scientific computing and information technology in ways that benefit both academic science and industry. Investments in improving chemical-based understanding and decision making will have a high impact because chemical science and engineering are at the foundation of a broad spectrum of technological and biological processes. If the United States is to maintain and strengthen its position as a world leader, chemical science and technology will have to aggressively pursue the opportunities offered by the advances in information and communication technologies.

At the intersection of information technology and the chemical sciences there are infrastructural challenges and opportunities. There are needs for infrastructure improvements that could enable chemical scientists and engineers to attain wholly new levels of computing-related research and education and demonstrate the value of these activities to society. These needs extend from research and teaching in the chemical sciences to issues associated with codes, software, data and storage, and networking and bandwidth.

Some things are currently working very well at the interface of computing and the chemical sciences. Networking and Internet high-speed connectivity have been integrated into the chemical sciences, changing the landscape of these fields and of computational chemistry. Commercial computational chemistry software companies and some academic centers provide and maintain computational and modeling codes that are widely used to solve problems in industry and academia. However, these companies and centers do not, and probably cannot, provide the infrastructure required for the development of new scientific approaches and codes for a research market that is deeply segmented. The development of new codes and applications by academia represents a mechanism for continuous innovation that drives the field and helps to direct the choice of application areas on which the power of computational chemistry and simulation is brought to bear. Modern algorithms and programming tools have speeded new code development and eased prototyping worries, but creating the complicated codes typical of chemical science and engineering applications remains an exceedingly difficult and time-consuming task. Defining new codes and applications is potentially a growth area of high value but one that faces major infrastructure implications if it is to be sustained.

Successful collaborations between chemists and chemical engineers, as well as broadly structured interdisciplinary groups in general, have grown rapidly during the past decade. These have created the demand for infrastructure development to solve important problems and new applications in ways never before envisioned. The current infrastructure must be improved if it is to be used effectively in interdisciplinary team efforts, especially for realizing the major potential impact of multiscale simulations. Infrastructure developments that support improved multidisciplinary interactions include resources for code development, assessment, and life-cycle maintenance; computers designed for science and engineering applications; and software for data collection, information management, visualization, and analysis. Such issues must be addressed broadly in the way that funding investments are made in infrastructure, as well as in cross-disciplinary education and in the academic reward structure.

The overarching infrastructure challenge is to provide at all times the needed accessibility, standardization and integration across platforms while also providing the fluidity needed to adapt to new concurrent advances in a time of rapid innovation.

RESEARCH

Significant gains in understanding and predictive ability are envisioned to result from the development of multiscale simulation methods for the investigation of complicated systems that encompass behavior over wide ranges of time and length scales. Such systems usually require a multidisciplinary approach. Often, multiscale simulations involve multiple phenomena that occur simultaneously with complex, subtle interactions that can confound intuition. While much is known about simulating aspects of behavior at individual scales (e.g., ab initio, stochastic, continuum, and supply-chain calculations), integration across scales is essential for understanding the behavior of entire systems.

A critical component in achieving the benefit implied by multiscale modeling will be funding for interdisciplinary research for which effective, collaborative web-based tools are required. The integration of computational results with experimental information is often necessary to solve multiscale problems. In some instances, creating opportunities to access shared equipment will be as critical as access to shared computers or software. Especially important is the ability to represent and understand the uncertainties, not only in the underlying scientific understanding, but also in experimental data that may come from extremely heterogeneous sources. The infrastructure to achieve these research goals must include definition of standard test cases for software and experiments.

Basic infrastructure needs include high-bandwidth access to high-performance computational facilities, further increased network and bus speed, diverse computer architectures, shared instruments, software, federated databases, storage, analysis, and visualization. Computers designed with a knowledge of the memory usage patterns of science and engineering problems will be useful, as will algorithms that take full advantage of the new generation of supercomputers. Continuation of the successful trend towards clusters of commodity computers may result in further opportunities for improved computational efficiency and cost effectiveness. Software should be characterized by interoperability and portability so that codes and computers can talk to each other and can be moved in a seamless manner to new systems when they become available.

EDUCATION

The need for student learning in basic mathematics at the intersection of computing and the chemical sciences is essential because it provides the foundation for computational chemistry, modeling, and simulation as well as associated software engineering. Although many entry-level students in the chemical sciences are familiar with the use of computers and programs, they often have little or no understanding of the concepts and design principles underlying their use. The integration of these topics in interdisciplinary courses is essential for the development of a skilled workforce.[1] Educational activities will require the investment of time and resources to develop new content in the curriculum for chemists and chemical engineers. New pedagogical approaches at both the undergraduate and graduate levels will be needed to address subjects at the interface of disciplines linked by scientific data, programming, and applications areas. Training students to adopt a problem-solving approach is critically important for good

---

[1]*Building a Workforce for the Information Economy,* National Research Council, National Academy Press, Washington, DC, 2001.

software engineering—and especially for producing codes and data structures that are useful to other people. A national community of educational open-source software would help speed development of training tools.

Just as training in mathematics and physics has been needed for work in chemical sciences and engineering, so will specific education in the use of modern IT tools, software design, and data structures be needed by the chemical professional of the twenty-first century. Such education will help in the rapid development of new approaches, cross-disciplinary integration, and integrated data handling and utilization.

Interdisciplinary research and development at the IT–chemical science interface are areas of great excitement and opportunity. Nevertheless, people trained to carry out such projects are in short supply. The continued capability of individuals requires both deep competence and the ability to interact across disciplines. The emphasis in graduate training therefore must be balanced between specialization within a discipline and cross-disciplinary collaboration and teamwork. Transfer of information between fields remains difficult when evaluating performance, particularly for tenure and promotion of faculty who focus on interdisciplinary projects or hold joint appointments in multiple departments. Such evaluation of scholarship will require attentive administrative coordination to resolve cultural differences. Creating high-quality educational programs to train people to work at interdisciplinary interfaces is currently a rate-limiting step in the growth of the field. Recognizing and rewarding the success of interdisciplinary scientists at different stages in their careers is becoming critically important for the sustained development of the field.

Computational chemistry and simulation methods should be incorporated into a broad range of educational programs to provide better understanding of the scope and limitations of various methods, as well as to facilitate their application over the full range of interdisciplinary problems to which they apply. Both science and engineering applications have to be addressed, since these can have different goals and methods of pursuit—with widely differing levels of sophistication. These include simple applications that can be helpful in early stages, complicated applications that require greater skill, and applications to truly complex nonlinear systems that represent the current focus of many experts in the field. Such training will benefit industry, where there is a need for computational specialists who understand the goals and objectives of a broad interdisciplinary problem and know how and when computational chemistry and systems-level modeling can provide added value. In academia, infrastructure support to facilitate better communication and interaction between chemists and chemical engineers will enhance the training of computational experts of the future. The field will be well served by establishing commonality in understanding and language between the creators and users of codes as well as the skilled computer science and engineering nonusers who develop the IT methods.

An increasingly important part of the infrastructure will be the skilled workers who maintain codes, software, and databases over their life cycle. The wide variety of tasks that require sustained management may necessitate a combination of local (funded through research grants) and national (funded through center grants) support to address the overall portfolio of needs.

Advances in the chemical sciences have permitted major advances in medicine, life science, earth science, physics and engineering, and environmental science, to name a few. Advances in productivity, quality of life, security, and economic vitality of global and American society have flowed directly from the efforts of people who work in these fields. Looking to the future, we need to build on these advances so that computational discovery and design can become standard components of broad education and training goals in our society. In this way, the human resources will be available to create, as well as to realize and embrace, the capabilities, challenges, and opportunities provided by the chemical sciences through advanced information technology.

Information and communication, data and informatics, and modeling and computing must become primary training goals for researchers in chemical science. These skills have to be accessible to effectively serve others in the society—from doctors to druggists, ecologists to farmers, and journalists to decision makers—who need an awareness of chemical phenomena to work effectively and to make wise decisions. Such skills provide liberating capabilities that enable interactions among people and facilitate new modes of thought, totally new capabilities for problem-solving, and new ways to extend the vision of the chemical profession and of the society it serves.

CODES, SOFTWARE, DATA AND BANDWIDTH

A critical issue for codes, software, and databases is maintenance over the life cycle of use. In the academic world, software with much potential utility can be lost with the graduation of students who develop the codes. Moreover, as codes become more complicated, the educational value of writing one's own code must be balanced against the nontrivial effort to move from a complicated idea, to algorithm, and then to code. Increasing fractions of student researchers are tending to develop skills with simpler practice codes, and then to work with and modify legacy codes that are passed down. Yet at the same time, working in a big coding environment with codes written by people who have long gone is difficult and often frustrating. Development of software that uses source-code generation to automatically fix dusty decks will be increasingly important for decreasing the time and effort associated with extending the useful life of codes. Also, the development of semiautomatic methods for generation of improved graphical user interfaces will reduce a significant barrier to sustaining the use of older code. Although the open-source approach works well for communities where thousands of eyes help remove bugs, it is unable to accommodate certain applications—for example, when proprietary information is involved.

Growth in multiscale simulation may be expected to drive development of improved tools for integration of different software systems, integration of different hardware architectures, and use of shared code by distributed collaborators. An increasing need will steadily result for improved interoperability and portability and for users to be able to modify codes created by others. Advances in object-oriented programming and component technology will help. Examples such as the Portable, Extensible Toolkit for Scientific Computing (PETSc) Library at Argonne National Laboratory represent the kind of infrastructure that will support growth in strategic directions.

Central to the vision of a continuously evolving code resource for new applications is the ability to build on existing concepts and codes that have been extensively developed. However, at present, academic code sharing and support mechanisms can at best be described as poor—sometimes as a result of perceived commercialization potential or competitive advantage. Moreover, code development and support are not explicitly supported by most research grants, nor is maintenance of legacy codes. Consequently, adapting academic codes from elsewhere may generate a risk that the code will become unsupported during its useful life cycle. Avoiding this risk results in continual duplication of effort—to produce trivial codes that could be better served by open-source toolkits and libraries maintained as part of the infrastructure.

The assurance of code verification, reliability, standardization, availability, maintenance, and security represents an infrastructure issue with broad implications. Sometimes commercial software has established a strong technical base, excellent interfaces, and user-friendly approaches that attract a wide range of users. Commercial software can be valuable when market forces result in continuous improvements that are introduced in a seamless manner, but generally, commercial code development is not well matched to the needs of small groups of research experts nor to many large potential markets of nonexperts. Therefore a critical need exists to establish standards and responsibilities for code.

The rapid growth of data storage per unit cost has been accompanied by equally significant increases in the demand for data, with the result that there is rapid increase in emphasis on data issues across chemical science and engineering. Bioinformatics and pharmaceutical database mining represent areas in which sophisticated methods have been effective in extracting useful knowledge from data. Newly emerging applications include scientific measurements, sensors in the environment, process-engineering data, manufacturing execution, and supply-chain systems. The overall goal is to build data repositories that can be accessed easily by remote computers to facilitate the use of shared data among creative laboratory scientists, plant engineers, process-control systems, business managers, and decision makers. Achieving this requires improved procedures that provide interoperability and data-exchange standards.

The integration of federated databases with predictive modeling and simulation tools represents an important opportunity for major advances in the effective use of massive amounts of data. The framework will need to include computational tools, evaluated experimental data, active databases, and knowledge-based software guides for generating chemical and physical property data on demand with quantitative measures of uncertainty. The approach has to provide validated, predictive simulation methods for complicated systems with seamless multiscale and multidisciplinary integration to predict properties and to model physical phenomena and processes. The results must be in a form that can be visualized and used by even a nonexpert.

In addition to the insightful use of existing data, the acquisition of new chemical and physical property data continues to grow in importance—as does the need to retrieve data for future needs. Such efforts require careful experimental measurements as well as skilled evaluation of related data from multiple sources. It will be necessary to assess confidence with robust uncertainty estimates; validate data with experimentally or calculated benchmark data of known accuracy; and document the metadata needed for interpretation.

There is a need to advance IT systems to provide scientific data and available bandwidth in the public arena. High-quality data represent the foundation upon which public and proprietary institutions can develop their knowledge-management and predictive modeling systems. It is appropriate that federal agencies participate in the growing number of data issues that are facing the chemical science and engineering community—including policy issues associated with access to data. Improved access to data not only will benefit research and technology but will provide policy and decision makers with superior insights on chemical data-centric matters such as environmental policy, natural resource utilization, and management of unnatural substances. Expanded bandwidth is crucial for collaborations, data flow and management, and shared computing resources.

> You might ask "What is the twenty-first century Grid infrastructure that is emerging?" I would answer that it is this tightly optically coupled set of data clusters for computing and visualization tied together in a collaborative middle layer. ... So, if you thought you had seen an explosion on the Internet, you really haven't seen anything yet.
>
> *Larry Smarr (Appendix D)*

ANTICIPATED BENEFITS OF INVESTMENT IN INFRASTRUCTURE

Chemical science and engineering serve major sectors that, in turn, have a wide range of expectations from infrastructure investments. At the heart of these is the development and informed use of data and simulation tools. The use of information technology to facilitate multidisciplinary teams that collaborate on large problems is in its infancy. Sustained investment in information technologies that facilitate the process of discovery and technological innovation holds truly significant promise, and the chemical sciences provide a large number of potential testbeds for the development of such capabilities.

In science and engineering research, the complex areas identified in Chapter 4 are clear points of entry for computer science, engineering, and applied mathematics along with chemical science and engineering. One of the great values of simulation is the insight it gives into the inner relationships of complicated systems—as well as the influence this insight has on the resulting outcome. The key enabling infrastructure elements are those that enhance the new intuitions and insights that are the first steps toward discovery.

> The advances being made in Grid technologies and virtual laboratories will enhance our ability to access and use computers, chemical data, and first-of-a-kind or one-of-a-kind instruments to advance chemical science and technology. Grid technologies will substantially reduce the barrier to using computational models to investigate chemical phenomena and to integrating data from various sources into the models or investigations. Virtual laboratories have already proven to be an effective means of dealing with the rising costs of forefront instruments for chemical research by providing capabilities needed by researchers not co-located with the instruments—all we need is a sponsor willing to push this technology forward on behalf of the user community.
> The twenty-first Century will indeed be an exciting time for chemical science and technology.
> *Thom Dunning (Appendix D)*

In industrial applications, tools are needed that speed targeted design and impact business outcomes through efficient movement from discovery to technological application. Valuing IT infrastructure tools requires recognizing how they enhance productivity, facilitate teamwork, and speed time-consuming experimental work.

Finding: Federal research support for individual investigators and for curiosity-driven research is crucial for advances in basic theory, formalisms, methods, applications, and understanding.

History shows that the investment in long-term, high-risk research in the chemical sciences must be maintained to ensure continued R&D progress that provides the nation's technological and economic well-being. Large-scale, large-group efforts are complementary to individual investigator

> ### Computer-Aided Design of Pharmaceuticals
>
> Computer-aided molecular design in the pharmaceutical industry is an application area that has evolved over the past several decades. Documentation of success in the pharmaceutical discovery process now transcends reports of individual applications of various techniques that have been used in a specific drug discovery program. The chemistry concepts of molecular size, shape, and properties and their influence on molecular recognition by receptors of complementary size, shape, and properties are central unifying concepts for the industry. These concepts and computational chemistry visualization tools are now used at will—and without hesitation—by virtually all participants, regardless of their core discipline (chemistry, biology, marketing, business, management). Such ubiquitous use of simple chemical concepts is an exceedingly reliable indicator of their influence and acceptance within an industry. The concepts that unify thinking in the pharmaceutical discovery field seemingly derive little from the complexity and rigor of the underlying computational chemistry techniques. Nevertheless, there is little reason to assume that these simple concepts could ever have assumed a central role without the support of computational chemistry foundations. In other words, having a good idea in science, or in industry, does not mean that anyone will agree with you (much less act on it) unless there is a foundation upon which to build.
>
> *Organizing Committee*

projects—both are crucial, and both are critically dependent on next-generation IT infrastructure.

Finding: A strong infrastructure at the intersection with information technology will be critical for the success of the nation's research investment in chemical science and technology.
The infrastructure includes hardware, computing facilities, research support, communications links, and educational structures. Infrastructure enhancements will provide substantial advantages in the pursuit of teaching, research, and development. Chemists and chemical engineers will need to be ready to take full advantage of capabilities that are increasing exponentially.

To accomplish this we must do the following:

- Recognize that significant investments in infrastructure will be necessary for progress.
- Enhance training throughout the educational system (elementary through

postgraduate) for computational approaches to the physical world. Assuring that chemists and chemical engineers have adequate training in information technology is crucial. Programming languages have been the traditional focus of such education; data structures, graphics, and software design are at least as important and should be an integral component (along with such traditional fundamental enablers as mathematics, physics, and biology) of the education of all workers in chemistry and chemical engineering.

• Maintain national computing laboratories with staff to support research users in a manner analogous to that for other user facilities.[2]

• Develop a mechanism to establish standards and responsibilities for verification, standardization, availability, maintenance, and security of codes.

• Define appropriate roles for developers of academic or commercial software throughout its life cycle.

• Provide universal availability of reliable and verified software.

The findings and recommendations outlined here and in previous chapters show that the intersection of chemistry and chemical engineering with computing and information technology is a sector that is ripe with opportunity. Important accomplishments have already been realized, and major technical progress should be expected if new and existing resources are optimized in support of research, education, and infrastructure. While this report identifies many needs and opportunities, the path forward is not yet fully defined and will require additional analysis.

Recommendation: Federal agencies, in cooperation with the chemical sciences and information technology communities, will need to carry out a comprehensive assessment of the chemical sciences–information technology infrastructure portfolio.

The information provided by such an assessment will provide federal funding agencies with a sound basis for planning their future investments in both disciplinary and cross-disciplinary research.

The following are among the actions that need to be taken:

• Identify criteria and appropriate indicators for setting priorities for infrastructure investments that promote healthy science and facilitate the rapid movement of concepts into well-engineered technological applications.

• Address the issue of standardization and accuracy of codes and databases, including the possibility of a specific structure or mechanism (e.g., within a federal laboratory) to provide responsibility for standards evaluation.

---

[2]*Cooperative Stewardship: Managing the Nation's Multidisciplinary User Facilities for Research with Synchrotron Radiation, Neutrons, and High Magnetic Fields,* National Research Council, National Academy Press, Washington, D.C., 1999.

- Develop a strategy for involving the user community in testing and adopting new tools, integration, and standards development. Federal investment in IT architecture, standards, and applications are expected to scale with growth of the user base, but the user market is deeply segregated, and there may not yet be a defined user base for any specific investment.
- Determine how to optimize incentives within peer-reviewed grant programs for creation of high quality cross-disciplinary software.

> During the next 10 years, chemical science and engineering will be participating in a broad trend in the United States and across the world: we are moving toward a distributed cyberinfrastructure. The goal will be to provide a collaborative framework for individual investigators who want to work with each other or with industry on larger-scale projects that would be impossible for individual investigators working alone.
>
> Larry Smarr (Appendix D)

Recommendation. In order to take full advantage of the emerging Grid-based IT infrastructure, federal agencies—in cooperation with the chemical sciences and information technology communities—should consider establishing several collaborative data–modeling environments. By integrating software, interpretation, data, visualization, networking, and commodity computing, and using web services to ensure universal access, these collaborative environments could impact tremendously the value of IT for the chemical community. They are ideal structures for distributed learning, research, insight, and development on major issues confronting both the chemical community and the larger society.

Collaborative Modeling-Data Environments should be funded on a multiyear basis; should be organized to provide integrated, efficient, standardized, state-of-the-art software packages, commodity computing and interpretative schemes; and should provide open-source approaches (where appropriate), while maintaining security and privacy assurance.

This report should be seen in the context of the larger initiative, *Beyond the Molecular Frontier: Challenges for Chemistry and Chemical Engineering,*[3] as well as in the six accompanying reports on societal needs (of which this report is

---

[3]*Beyond the Molecular Frontier: Challenges for Chemistry and Chemical Engineering,* National Research Council, The National Academies Press, Washington, D.C., 2003.

one).[4,5,6,7,8] This component on *Information and Communications*, examines perhaps the most dynamically growing capability in the chemical sciences. The findings reported in the Executive Summary and in greater depth in the body of the text constitute what the committee believes to be viable and important guidance to help the chemical sciences community to take full advantage of growing IT capabilities for the advancement of the chemical sciences and technology—and thereby for the betterment of our society and our world.

---

[4]*Challenges for the Chemical Sciences in the 21st Century: National Security & Homeland Defense,* National Research Council, The National Academies Press, Washington, D.C., 2002.

[5]*Challenges for the Chemical Sciences in the 21st Century: Materials Science and Technology,* National Research Council, The National Academies Press, Washington, D.C., 2003.

[6]*Challenges for the Chemical Sciences in the 21st Century: Energy and Transportation,* National Research Council, The National Academies Press, Washington, D.C., 2003 (in preparation).

[7]*Challenges for the Chemical Sciences in the 21st Century: The Environment,* National Research Council, The National Academies Press, Washington, D.C., 2003.

[8]*Challenges for the Chemical Sciences in the 21st Century: Health and Medicine,* National Research Council, The National Academies Press, Washington, D.C., 2003 (in preparation).

Appendixes

A

Statement of Task

The Workshop on Information and Communications is one of six workshops held as part of "Challenges for the Chemical Sciences in the 21st Century." The workshop topics reflect areas of societal need—materials and manufacturing, energy and transportation, national security and homeland defense, health and medicine, information and communications, and environment. The charge for each workshop was to address the four themes of discovery, interfaces, challenges, and infrastructure as they relate to the workshop topic:

- Discovery—major discoveries or advances in the chemical sciences during the last several decades
- Interfaces—interfaces that exist between chemistry/chemical engineering and such areas as biology, environmental science, materials science, medicine, and physics
- Challenges—the grand challenges that exist in the chemical sciences today
- Infrastructure—infrastructure that will be required to allow the potential of future advances in the chemical sciences to be realized

B

Biographies of the Organizing Committee Members

Richard C. Alkire (Co-Chair) is the Charles and Dorothy Prizer Professor of Chemical Engineering at the University of Illinois at Urbana-Champaign. He has recently served as vice chancellor for research and dean of the Graduate College. His current research is in the area of electrochemical deposition and dissolution of metals, including corrosion, and the strategic use of high-performance computing in collaborative problem solving. He received his B.S. from Lafayette College and Ph.D. from University of California at Berkeley. He is a member of the National Academy of Engineering.

Mark A. Ratner (Co-Chair) is professor of chemistry, Department of Chemistry, Northwestern University, Evanston, Illinois. He obtained his B.A. from Harvard University (1964) and his Ph.D. (chemistry) from Northwestern University (1969). His research interests are in nonlinear optical response properties of molecules, electron transfer and molecular electronics, dynamics of polymer electrolyte transport, self-consistent field models for coupled vibration reaction dynamics, mean-field models for extended systems, and tribology and glassy dynamics. He has some 312 professional publications. Ratner is a fellow of the American Physical Society and the American Association for the Advancement of Science and has received numerous teaching awards from Northwestern University. He is a member of the National Academy of Sciences.

Peter T. Cummings is the John R. Hall Professor of Chemical Engineering at Vanderbilt University. In addition, he is on the staff of the Chemical Sciences Division at Oak Ridge National Laboratory, where he also holds the position of director of the Nanomaterials Theory Institute of the Center for Nanophase Materials Science. He received his bachelor's degree in mathematics from the University of Newcastle (New South Wales, Australia) and his Ph.D. from the University of Melbourne (Victoria, Australia) in applied mathematics. His research expertise is in statistical thermodynamics, molecular modeling, computational science, and chemical process design.

Ignacio E. Grossmann (Steering Committee Liason) is Rudolph H. and Florence Dean Professor and head of the chemical engineering department at Carnegie Mellon University. He received his B.Sc. (1974) from Universidad Iberoamericana, Mexico, and his M.Sc. (1975) and Ph.D. (1977) degrees from Imperial College, London. He joined Carnegie Mellon in 1979 and has focused his research on the synthesis of integrated flow sheets, batch processes, and mixed-integer optimization. The goals of his work are to develop novel mathematical programming models and techniques in process systems engineering. He was elected to the Mexican Academy of Engineering in 1999, and he is a member of the National Academy of Engineering.

Judith C. Hempel was assistant chair of the Department of Biomedical Engineering at the University of Texas, Austin, 2002-2003. Previously, she was associate director of the Molecular Design Institute at the University of California, San Francisco, and was a member of the scientific staff of the computational chemistry software company Biosym/MSI and a founding member of the computer-aided drug design group at SmithKline. She received her B.S., M.A., and Ph.D. degrees in chemistry at the University of Texas, Austin. She served as a member of the Computer Science and Telecommunications Board from 1996 to 2001. Her research expertise is in theoretical and computational chemistry.

Kendall N. Houk is professor of chemistry in the Department of Chemistry and Biochemistry at the University of California, Los Angeles. He obtained his A.B. in 1964 and his Ph.D. in 1968, both from Harvard University. His research involves the use of computational methods for the solution of chemical and biological problems, and he continues experimental research as well. He has some 560 professional publications. He served on the Board on Chemical Sciences and Technology from 1991 to 1994. From 1988 to 1991 Houk was director of the NSF Chemistry Division. He was elected to the American Academy of Arts and Sciences in 2002.

Sangtae Kim (Board on Chemical Sciences and Technology liaison) is vice president and information officer of the Lilly Research Laboratories. Prior to Eli Lilly, he was a vice president and head of R&D information technology for the Parke-Davis Pharmaceutical Research division of Warner-Lambert Company. Prior to his positions in industry, he was the Wisconsin Distinguished Professor of Chemical Engineering at the University of Wisconsin-Madison. He received his B.S. and M.S. degrees from the California Institute of Technology (1979) and his Ph.D. degree in chemical engineering from Princeton University (1983). His research interests include computational fluid dynamics, computational biology and high-performance computing. He is a member of the National Academy of Engineering.

Kenny B. Lipkowitz is chairman and professor of chemistry at North Dakota State University. Prior to assuming this position in 2003, he was professor of chemistry at the Purdue University School of Science and associate director of chemical informatics in the Indiana University School of Informatics. He received his B.S. in chemistry from the State University of New York at Geneseo in 1972 and his Ph.D. in organic chemistry from Montana State University in 1975. Lipkowitz's work in informatics focuses on the derivation of three-dimensional descriptors of molecules for use in quantitative structure activity relationships (QSAR), one of several techniques used for computer-assisted molecular design. He is associate editor of the *Journal of Chemical Information and Computer Science,* and editor of *Reviews in Computational Chemistry.*

Julio M. Ottino is currently Robert R. McCormick Institute Professor and Walter P. Murphy Professor of Chemical Engineering at Northwestern University in Evanston, Illinois. He was chairman of the Department of Chemical Engineering from 1992 to 2000. Ottino received a Ph.D. in chemical engineering from the University of Minnesota and has held research positions at the California Institute of Technology and Stanford University. His research interests are in the area of complex systems and nonlinear dynamics with applications to fluids and granular matter. He is a fellow of the American Physical Society and the American Association for the Advancement of Science. He is a member of the National Academy of Engineering.

John C. Tully is Arthur T. Kemp Professor of Chemistry, Physics, and Applied Physics in the Department of Chemistry at Yale University. Tully is a leading theorist studying the dynamics of gas surface interactions. He develops theoretical and computational methods to address fundamental problems and then works with experimentalists to integrate theory with observation. Energy exchange and redistribution, adsorption and desorption, and dissociation and recombination are among surface phenomena he has elucidated. He uses mixed quantum-classical dynamics, which allow the extension of conventional molecular dynamics simulation methods to processes involving electronic transitions or quantum atomic motion. He is a member of the National Academy of Sciences.

Peter G. Wolynes (Steering Committee Liaison) is professor of chemistry and biochemistry at the University of California, San Diego. He was previously professor of chemistry at the University of Illinois at Urbana-Champaign. He received his A.B. from Indiana University in 1971 and his Ph.D. from Harvard University in 1976. His research area is physical chemistry with specialized interests in chemical physics of condensed matter, quantum dynamics and reaction kinetics in liquids, dynamics of complex fluids, phase transitions and the glassy state, and biophysical applications of statistical mechanics, especially protein folding. He is a member of the National Academy of Sciences.

C

Workshop Agenda

**Agenda
Workshop on Information and Communications
Challenges for the Chemical Sciences in the 21st Century**

National Academy of Sciences
2101 Constitution Avenue, NW
Washington, DC

THURSDAY, OCTOBER 31, 2002

   7:30   Breakfast

SESSION 1. OVERVIEW AND IMPACT
   8:00   Introductory remarks by organizers—Background of project
              DOUGLAS J. RABER, National Research Council
              RONALD BRESLOW, MATTHEW V. TIRRELL, Co-Chairs, *Steering Committee on Challenges for the Chemical Sciences in the 21st Century*
              RICHARD C. ALKIRE, MARK A. RATNER, Co-Chairs, Information & Communications Workshop Committee
   8:20   JAMES R. HEATH, *University of California, Los Angeles*
              The Current State of the Art in Nanoscale and Molecular Information Technologies
   8:50   DISCUSSION
   9:10   THOM H. DUNNING, JR., *Oak Ridge National Laboratory and University of Tennessee*
              Information & Communication Technologies and Chemical Science Technology

9:40 DISCUSSION

10:00 **BREAK**
10:30 JUAN J. DE PABLO, *University of Wisconsin, Madison*
The Evolution of Structure Modeling
11:00 DISCUSSION
11:20 CHARLES H. BENNETT, *IBM Research*
Quantum Information
11:50 DISCUSSION
12:10 LUNCH

SESSION 2. DATA AND INFORMATICS
1:30 CHRISTODOULOS A. FLOUDAS, *Princeton University*
Systems Approaches in Bioinformatics and Computational Genomics
2:00 DISCUSSION
2:20 ANNE M. CHAKA, *National Institute of Standards and Technology*
How Scientific Computing, Knowledge Management, and Databases Can Enable Advances and New Insights in Chemical Technology
2:50 DISCUSSION

3:10 BREAKOUT SESSION: DISCOVERY
What major discoveries or advances related to information and communications have been made in the chemical sciences during the last several decades?

4:15 **BREAK**
4:30 Reports from breakout sessions and discussion
5:30 RECEPTION
6:00 BANQUET
Speaker: LARRY L. SMARR, *University of California, San Diego*

FRIDAY, NOVEMBER 1, 2002
7:30 BREAKFAST

SESSION 3. SIMULATIONS AND MODELING (PART 1)
8:00 DENNIS J. UNDERWOOD, *Bristol-Myers Squibb Company*
Drug Discovery, a Game of 20 Questions
8:30 DISCUSSION
8:50 GEORGE C. SCHATZ, *Northwestern University*
Simulation in Materials Science
9:20 DISCUSSION

APPENDIX C 69

9:40   BREAKOUT SESSION: INTERFACES
       What are the major computing-related discoveries and challenges at the interfaces between chemistry/chemical engineering and other disciplines, including biology, environmental science, information science, materials science, and physics?

10:45 **BREAK**
11:00 Reports from breakout sessions and discussions
12:00 LUNCH

SESSION 4. SIMULATIONS AND MODELING (PART 2)
1:00   ELLEN B. STECHEL, *Ford Motor Company*
       Modeling and Simulation as a Design Tool
1:30   DISCUSSION
1:50   LINDA R. PETZOLD, *University of California, Santa Barbara*
       The Coming Age of Computational Sciences
2:20   DISCUSSION

2:40   BREAKOUT SESSION: CHALLENGES
       What are the information and communications grand challenges in the chemical sciences and engineering

3:45   **BREAK**
4:00   Reports from breakout sessions and discussion
5:00   ADJOURN FOR DAY

SATURDAY, NOVEMBER 2, 2002
7:30   BREAKFAST

SESSION 5. ACCESSIBILITY, STANDARDIZATION, AND INTEGRATION
8:00   DIMITRIOS MAROUDAS, *University of Massachusetts, Amherst*
       Multiscale Modeling
8:30   DISCUSSION
8:50   RICHARD FRIESNER, *Columbia University*
       Modeling of Complex Chemical Systems Relevant to Biology and Materials Science: Problems and Prospects
9:20   DISCUSSION

9:40   BREAKOUT SESSION: INFRASTRUCTURE
       What are the two issues at the intersection of computing and the chemical sciences for which there are structural challenges and opportunities—in teaching, research, equipment, codes and software, facilities, and personnel?

10:45	**BREAK**
11:00	Reports from breakout sessions (and discussion)
12:00	Wrap-up and closing remarks RICHARD C. ALKIRE, MARK A. RATNER, Co-Chairs, Information and Communications Workshop Committee
12:15	ADJOURN

D

Workshop Presentations

QUANTUM INFORMATION

Charles H. Bennett
IBM Research

This paper discussed how some fundamental ideas from the dawn of the last century have changed our understanding of the nature of information. The foundations of information processing were really laid in the middle of the twentieth century, and only recently have we become aware that they were constructed a bit wrong—that we should have gone back to the quantum theory of the early twentieth century for a more complete picture.

The word calculation comes from the Latin word meaning a pebble; we no longer think of pebbles when we think of computations. Information can be separated from any particular physical object and treated as a mathematical entity. We can then reduce all information to bits, and we can deal with processing these bits to reveal implicit truths that are present in the information. This notion is best thought of separately from any particular physical embodiment.

In microscopic systems, it is not always possible (and generally it is not possible) to observe a system without disturbing it. The phenomenon of entanglement also occurs, in which separated bodies could be correlated in a way that cannot be explained by traditional classical communication. The notion of quantum information can be abstracted, in much the same way as the notions of classical information. There are actually more things that can be done with information, if it is regarded in this quantum fashion.

The analogy between quantum and classical information is actually straightforward—classical information is like information in a book or on a stone tablet, but quantum information is like information in a dream: we try to recall the dream and describe it; each description resembles less the original dream than the previous description.

The first, and so far the most practical, application of quantum information theory is quantum cryptography. Here one encodes messages and passes them on. If one thinks of photon polarization, one can distinguish vertical and horizontal polarization through calcite crystals, but diagonal photons are in principle not reliably distinguishable. They should be thought of as a superposition of vertical and horizontal polarization, and they actually propagate containing aspects of both of these parent states. This is an entangled structure, and the entangled structure contains more information than either of the pure states of which it is composed.

The next step beyond quantum cryptography, the one that made quantum information theory a byword, is the discovery of fast quantum algorithms for solving certain problems. For quantum computing, unlike simple cryptography, it is necessary to consider not only the preparation and measurement of quantum states, but also the interaction of quantum data along a stream. This is technically a more difficult issue, but it gives rise to quite exciting basic science involving quantum computing.

The entanglement of different state structures leads to so-called Einstein-Podolsky-Rosen states, which are largely responsible for the unusual properties of quantum information.

The most remarkable advance in the field, the one that made the field famous, is the fast factor algorithm discovered by Shor at Bell Laboratories. It demonstrates that exponential speedup can be obtained using a quantum computer to factor large numbers into their prime components. Effectively, this quantum factorization algorithm works because it is no more difficult, using a quantum computer, to factor a large number into its prime factors than it is to multiply the prime factors to produce the large number. It is this condition that renders a quantum computer exponentially better than a classical computer in problems of this type.

The above considerations deal with information in a disembodied form. If one actually wants to make a quantum computer, there are all sorts of fabrication, interaction, decoherence, and interference considerations. This is a very rich area of experimental science, and many different avenues have been attempted. Nuclear magnetic resonance, ion traps, molecular vibrational states, and solid-state implementations have all been used in attempts to produce actual quantum computers.

Although an effective experimental example of a quantum computer is still incomplete, many major theoretical advances have suggested that some of the obvious difficulties can in fact be overcome. Several discoveries indicate that the effects of decoherence can be prevented, in principle, by including quantum error correcting codes, entanglement distillation, and quantum fault-tolerant circuits. Good quantum computing hardware does not yet exist. The existence of these methodologies means that the hardware does not have to be perfectly efficient or perfectly reliable, because these programming techniques can make arbitrarily good quantum computers possible, even with physical equipment that suffers from decoherence issues.

Although most of the focus has been on quantum cryptography, quantum processing provides an important example of the fact that quantum computers not only can do certain tasks better than ordinary computers, but also can do different tasks that would not be imagined in the context of ordinary information processing. For example, entanglement can enhance the communication of classical messages, by augmenting the capacity of a quantum channel for sending messages.

To summarize, quantum information obeys laws that subtly extend those governing classical information. The way in which these laws are extended is reminiscent of the transition from real to complex numbers. Real numbers can be viewed as an interesting subset of complex numbers, and some questions that might be asked about real numbers can be most easily understood by utilization of the complex plane. Similarly, some computations involving real input or real output (by "real" I mean classical) are most rapidly developed using quantum intermediate space in quantum computers. When I'm feeling especially healthy, I say that quantum computers will probably be practical within my lifetime—strange phenomena involving quantum information are continually being discovered.

HOW SCIENTIFIC COMPUTING KNOWLEDGE MANAGEMENT AND DATABASES CAN ENABLE ADVANCES AND NEW INSIGHTS IN CHEMICAL TECHNOLOGY

Anne M. Chaka
National Institute of Standards and Technology

The focus of this paper is on how scientific computing and information technology (IT) can enable technical decision making in the chemical industry. The paper contains a current assessment of scientific computing and IT, a vision of where we need to be, and a roadmap of how to get there. The information and perspectives presented here come from a wide variety of sources. A general perspective from the chemical industry is found in the Council on Chemical Research's Vision 2020 Technology Partnership,[1] several workshops sponsored

---

[1] *Chemical Industry of the Future: Technology Roadmap for Computational Chemistry*, Thompson, T. B., Ed., Council for Chemical Research, Washington, DC, 1999; *http://www.ccrhq.org/vision/index/*

by NSF, NIST,[2] and DOE, and the WTEC report on industrial applications of molecular and materials modeling (which contains detailed reports on 91 institutions, including over 75 chemical companies, plus additional data from 55 U.S. chemical companies and 256 world-wide institutions (industry, academia, and government).[3] My own industrial perspective comes from my position as co-leader for the Lubrizol R&D IT Vision team for two years, and ten years as the head of computational chemistry and physics prior to coming to NIST. It should be noted that although this paper focuses primarily on the chemical industry, many of the same issues apply to the biotechnology and materials industry.

There are many factors driving the need for scientific computing and knowledge management in the chemical industry. Global competition is forcing U.S. industry to reduce R&D costs and the time to develop new products in the chemical and materials sectors. Discovery and process optimization are currently limited by a lack of property data and insight into mechanisms that determine performance. Thirty years ago there was a shortage of chemicals, and customers would pay premium prices for any chemical that worked at all. Trial and error was used with success to develop new chemistry. Today, however, the trend has shifted due to increased competition from an abundance of chemicals that work on the market, customer consolidation, and global competition that is driving commodity pricing even for high-performance and fine chemicals. Trial and error have become too costly, and the probability of success is too low. Hence it is becoming widely recognized that companies need to develop and fine-tune chemicals and formulations by design in order to remain competitive, and to screen chemicals prior to a long and costly synthesis and testing process. In addition, the chemicals produced today must be manufactured in a way that minimizes pollution and energy costs. Higher throughput is being achieved by shifting from batch processing to continuous feed stream, but this practice necessitates a greater understanding of the reaction kinetics, and hence the mechanism, to optimize feed stream rates. Simulation models are needed that have sufficient accuracy to be able to predict what upstream change in process variables are required to maintain the downstream products within specifications, as it may take several hours for upstream changes to affect downstream quality. Lastly, corporate downsizing is driving the need to capture technical knowledge in a form that can be queried and augmented in the future.

---

[2]NIST Workshop on Predicting Thermophysical Properties of Industrial Fluids by Molecular Simulations (June, 2001), Gaithersburg, MD; 1st International Conference on Foundations of Molecular Modeling and Simulation 2000: Applications for Industry (July, 2000), Keystone, CO; Workshop on Polarizability in Force Fields for Biological Simulations (December 13-15, 2001), Snowbird, UT.

[3]*Applying Molecular and Materials Modeling*, Westmoreland, P. R.; Kollman, P. A.; Chaka, A. M.; Cummings, P. T.; Morokuma, K.; Neurock, M.; Stechel, E.B .; Vashishta, P., Eds. Kluwer Academic Publishers, Dordrecht, 2002; *http://www.wtec.org/*.

Data and property information are most likely to be available on commodity materials, but industrial competition requires fast and flexible means to obtain data on novel materials, mixtures, and formulations under a wide range of conditions. For the vast majority of applications, particularly those involving mixtures and complex systems (such as drug-protein interactions or polymer nanocomposites), evaluated property data simply do not exist and are difficult, time consuming, or expensive to obtain. For example, measuring the density of a pure liquid to 0.01% accuracy requires a dual sinker apparatus costing $500,000, and approximately $10,000 per sample. Commercial laboratory rates for measuring vapor-liquid equilibria for two state points of a binary mixture are on the order of $30,000 to 40,000. Hence industry is looking for a way to supply massive amounts of data with reliable uncertainty limits *on demand*. Predictive modeling and simulation have the potential to help meet this demand.

Scientific computing and information technology, however, have the potential to offer so much more than simply calculating properties or storing data. They are essential to the organization and transformation of data into wisdom that enables better technical decision making. The information management pyramid can be assigned four levels defined as follows:

1. *Data:* a disorganized, isolated set of facts
2. *Information:* organized data that leads to insights regarding relationships—knowing *what* works
3. *Knowledge:* knowing *why* something works
4. *Wisdom:* having sufficient understanding of factors governing performance to reliably predict what will happen—knowing what *will* work

To illustrate how scientific computing and knowledge management convert data and information into knowledge and wisdom, a real example is taken from lubricant chemistry. Polysulfides, $R\text{-}S^n\text{-}R$, are added to lubricants to prevent wear of ferrous metal components under high pressure. The length of the polysulfide chain, n, is typically between 2 and 6. A significant performance problem is that some polysulfide formulations also cause corrosion of copper-containing components such as bronze or brass. To address this problem, a researcher assembles data from a wide variety of sources such as analytical results regarding composition, corrosion and antiwear performance tests, and field testing. IT makes it easy for the right facts to be gathered, visualized, and interpreted. After analysis of the data, the researcher comes to the realization that long-chain polysulfides ($n = 4$ or greater) corrode copper, but shorter chains ($n = 2$ to 3) do not. This is knowing *what* happens. Scientific computing and modeling can then be used to determine *why* something happens. In this case, quantum mechanics enabled us to understand that the difference in reactivity of these sulfur chains could be explained by significant stabilization of the thiyl radical delocalized over two adjacent sulfur atoms after homolytic cleavage of the S-S bond: R-SS-SS-R $\to$ 2R-SS•. The monosulfur thiyl radical R-S• was significantly higher in energy and therefore is much less likely to form. Hence copper corrosion is consistent with the formation of stable thiyl radicals. This insight led to a generalization that almost any sulfur radical with a low energy of formation will likely corrode copper, and we were able to reliably predict copper corrosion performance from the chemical structure of a sulfur-containing species prior to testing. This understanding also led to improvements in the manufacturing process and other applications of sulfur chemistry, and is an example of what is meant by wisdom (i.e., reliably predicting what *will* happen in novel applications due to a fundamental understanding of the underlying chemistry and physics).

What is the current status of scientific computing and knowledge management with respect to enabling better technical decisions? For the near term, databases, knowledge management, and scientific computing are currently most effective when they enable human insight. We are a long way from hitting a carriage return and obtaining answers to tough problems automatically, if ever. Wetware (i.e., human insight) is currently the best link between the levels of data, information, knowledge, and wisdom. There is no substitute for critical, scientific thinking. We can, however, currently expect an idea to be tested via experiment or calculation. First principles calculations, if feasible on the system, improve the robustness of the predictions and can provide a link between legacy data and novel chemistry applications. Computational and IT methods can be used to generate a set of possibilities combinatorially, analyze the results for trends, and visualize the data in a manner that enables scientific insight. Developing these systems is resource intensive and very application specific. Companies will invest in their development for only the highest priority applications, and the knowledge gained will be proprietary. Access to data is critical for academics in the QSPR-QSAR method development community, but is problematic due to intellectual property issues in the commercial sector.[4] Hence there is a need to advance the science and the IT systems in the public arena to develop the fundamental foundation and building blocks upon which public and proprietary institutions can develop their own knowledge management and predictive modeling systems.

What is the current status of chemical and physical property data? Published evaluated chemical and physical property data double every 10 years, yet this is woefully inadequate to keep up with demand. Obtaining these data requires meticulous experimental measurements and/or thorough evaluations of related data from multiple sources. In addition, data acquisition processes are time- and resource-consuming and therefore must be initiated well in advance of an anticipated need within an industrial or scientific application. Unfortunately a significant part of the existing data infrastructure is not directly used in any meaningful

---

[4]Comment by Professor Curt Breneman, Rensselaer Polytechnic Institute.

application because data requirements often shift between the initiation and completion of a data project. Analysis and fitting, such as for equation-of-state models, must be reinitiated when significant new data become available.

One vision that has been developed in consultation with the chemical and materials industries can be described as a "Universal Data and Simulation Engine." This engine is a framework of computational tools, evaluated experimental data, active databases, and knowledge-based software guides for generating chemical and physical property data *on demand* with quantitative measures of uncertainty. This engine provides validated, predictive simulation methods for complex systems with seamless multiscale and multidisciplinary integration to predict properties and model physical phenomena and processes. The results are then visualized in a form useful for scientific interpretation, sometimes by a non-expert. Examples of high-priority challenges cited by industry in the WTEC report to be ultimately addressed by the Universal Data and Simulation Engine are discussed below.[5]

How do we achieve this vision of a Universal Data and Simulation Engine? Toward this end, NIST has been exploring the concepts of dynamic data evaluation and virtual measurements of chemical and physical properties and predictive simulations of physical phenomena and processes. In dynamic data evaluation, all available experimental data within a technical area are collected routinely and continuously, and evaluations are conducted dynamically—using an automated system—when information is required. The value of data is directly related to the uncertainty, so "recommended" data must include a robust uncertainty estimate. Metadata are also collected (i.e., auxiliary information required to interpret the data such as experimental method). Achieving this requires interoperability and data exchange standards. Ideally the dynamic data evaluation is supplemented by calculated data based on validated predictive methods (virtual measurements), and coupled with a carefully considered experimental program to generate benchmark data.

Both virtual measurements and the simulation engine have the potential to meet a growing fraction of this need by supplementing experiment and providing data in a timely manner at lower cost. Here we define "virtual measurements" specifically as predictive modeling tools that yield property data with quantified uncertainties analogous to observable quantities measured by experiment (e.g., rate constants, solubility, density, and vapor-liquid equilibria). By "simulation" we mean validated modeling of processes or phenomena that provides insight

---

[5]These include liquid-liquid interfaces (micelles and emulsions), liquid-solid interfaces (corrosion, bonding, surface wetting, transfer of electrons and atoms from one phase to another), chemical and physical vapor deposition (semiconductor industry, coatings), and influence of chemistry on the thermomechanical properties of materials, particularly defect dislocation in metal alloys; complex reactions in multiple phases over multiple time scales. Solution properties of complex solvents and mixtures (suspending asphaltenes or soot in oil, polyelectrolytes, free energy of solvation rheology), composites (nonlinear mechanics, fracture mechanics), metal alloys, and ceramics.

into mechanisms of action and performance with atomic resolution that is not directly accessible by experiment but is essential to guide technical decision making in product design and problem solving. This is particularly crucial for condensed-phase processes where laboratory measurements are often the average of myriad atomistic processes and local conditions that cannot be individually resolved and analyzed by experimental techniques. It is analogous to gas-phase kinetics in the 1920s prior to modern spectroscopy when total pressure was the only measurement possible. The foundation for virtual measurements and simulations is experimental data and mathematical models that capture the underlying physics at the required accuracy of a given application. Validation of theoretical methods is vitally important.

The Council for Chemical Research's Vision 2020 states that the desired target characteristics for a virtual measurement system for chemical and physical properties are as follows: problem setup requires less than two hours, completion time is less than two days, cost including labor is less than $1,000 per simulation, and it is usable by a nonspecialist (i.e., someone who cannot make a full-time career out of molecular simulation). Unfortunately, we are a long way from meeting this target, particularly in the area of molecular simulations. Quantum chemistry methods have achieved the greatest level of robustness and—coupled with advances in computational speed—have enabled widespread success in areas such as predicting gas-phase, small-molecule thermochemistry and providing insight into reaction mechanisms. Current challenges for quantum chemistry are accurate predictions of rate constants and reaction barriers, condensed-phase thermochemistry and kinetics, van der Waals forces, searching a complex reaction space, transition metal and inorganic systems, and performance of alloys and materials dependent upon the chemical composition.

A measure of the current value of quantum mechanics to the scientific community is found in the usage of the NIST Computational Chemistry Comparison and Benchmark Database (CCCBDB), (*http://srdata.nist.gov/cccbdb*). The CCCBDB was established in 1997 as a result of an (American Chemical Society) ACS workshop to answer the question, How good is that ab initio calculation? The purpose is to expand the applicability of computational thermochemistry by providing benchmark data for evaluating theoretical methods and assigning uncertainties to computational predictions. The database contains over 65,000 calculations on 615 chemical species for which there are evaluated thermochemical data. In addition to thermochemistry, the database also contains results on structure, dipole moments, polarizability, transition states, barriers to internal rotation, atomic charges, etc. Tutorials are provided to aid the user in interpreting data and evaluating methodologies. Since the CCCBDB's inception, usage has doubled every year up to the current sustained average of 18,000 web pages served per month, with a peak of over 50,000 pages per month. Last year over 10,000 separate sites accessed the CCCBDB. There are over 400 requests per month for new chemical species not contained in the database.

The CCCBDB is currently the only computational chemistry or physics database of its kind. This is due to the maturity of quantum mechanics to reliably predict gas-phase thermochemistry for small (20 nonhydrogen atoms or less), primarily organic, molecules, plus the availability of standard-reference-quality experimental data. For gas-phase kinetics, however, only in the past two years have high-quality (<2% precision) rate-constant data become available for H• and •OH transfer reactions to begin quantifying uncertainty for the quantum mechanical calculation of reaction barriers and tunneling.[6] There is a critical need for comparable quality rate data and theoretical validation for a broader class of gas-phase reactions, as well as solution phase for chemical processing and life science, and surface chemistry.

One of the highest priority challenges for scientific computing for the chemical industry is the reliable prediction of fluid properties such as density, vapor-liquid equilibria, critical points, viscosity, and solubility for process design. Empirical models used in industry have been very useful for interpolating experimental data within very narrow ranges of conditions, but they cannot be extended to new systems or to conditions for which they were not developed. Models based exclusively on first principles are flexible and extensible, but can be applied only to very small systems and must be "coarse-grained" (approximated by averaging over larger regions) for the time and length scales required in industrial applications. Making the connection between quantum calculations of binary interactions or small clusters and properties of bulk systems (particularly systems that exhibit high-entropy or long-range correlated behavior) requires significant breakthroughs and expertise from multiple disciplines. The outcome of the First Industrial Fluid Properties Simulation Challenge[7] (sponsored by AIChE's Computational Molecular Science and Engineering Forum and administered by NIST) underscored these difficulties and illustrated how fragmented current approaches are. In industry, there have been several successes in applying molecular simulations, particularly in understanding polymer properties, and certain direct phase equilibrium calculations. Predicting fluid properties via molecular simulation, however, remains an art form rather than a tool. For example, there are currently over a dozen popular models for water, but models that are parameterized to give good structure for the liquid phase give poor results for ice. Others, parameter-

---

[6]Louis, F.; Gonzalez, C.; Huie, R. E.; Kurylo, M. J. *J. Phys. Chem. A* **2001**, *105*, 1599-1604.

[7]The goals of this challenge were to: (a) obtain an in-depth and objective assessment of our current abilities and inabilities to predict thermophysical properties of industrially challenging fluids using computer simulations, and (b) drive development of molecular simulation methodology toward a closer alignment with the needs of the chemical industry. The Challenge was administered by NIST and sponsored by the Computational Molecular Science and Engineering Forum (AIChE). Industry awarded cash prizes to the champions to each of the three problems (vapor-liquid equilibrium, density, and viscosity). Results were announced at the AIChE annual meeting in Indianapolis, IN, November 3, 2002.

ized for solvation and biological applications, fail when applied to vapor-liquid equilibrium properties for process engineering. The lack of "transferability" of water models indicates that the underlying physics of the intermolecular interactions is not adequately incorporated. The tools and systematic protocols to customize and validate potentials for given properties with specified accuracy and uncertainty do not currently exist and need to be developed.

In conclusion, we are still at the early stages of taking advantage of the full potential offered by scientific computing and information technology to benefit *both* academic science and industry. A significant investment is required to advance the science and associated computational algorithms and technology. The impact and value of improving chemical-based insight and decision making are high, however, because chemistry is at the foundation of a broad spectrum of technology and biological processes such as

- how a drug binds to an enzyme,
- manufacture of semiconductors,
- chemical reactions occurring inside a plastic that makes it burn faster than others, and
- how defects migrate under stress in a steel I-beam.

A virtual measurement system can serve as a framework for coordinating and catalyzing academic and government laboratory science in a form useful for solving technical problems and obtaining properties. There are many barriers to obtaining the required datasets that must be overcome, however. Corporate data are largely proprietary, and in academia, generating data is "perceived as dull so it doesn't get funded."[8] According to Dr. Carol Handwerker (chief, Metallurgy Division, Materials Science and Engineering Laboratory, NIST), even at NIST just gathering datasets in general is not well supported at the moment, because it is difficult for NIST to track the impact that the work will have on the people who are using the datasets. One possible way to overcome this barrier may be to develop a series of standard test problems in important application areas where the value can be clearly seen. The experimental datasets would be collected and theoretical and scientific computing algorithms would be developed, integrated, and focused in a sustained manner to move all the way through the test problems. The data collection and scientific theory and algorithm development then clearly become means to an end.

---

[8]Comment by Professor John Tully, Yale University.

ON THE STRUCTURE OF MOLECULAR MODELING: MONTE CARLO METHODS AND MULTISCALE MODELING

Juan J. de Pablo
University of Wisconsin, Madison

The theoretical and computational modeling of fluids and materials can be broadly classified into three categories, namely atomistic or molecular, mesoscopic, and continuum or macroscopic.[1] At the atomistic or molecular level, detailed models of the system are employed in molecular simulations to predict the structural, thermodynamic, and dynamic behavior of a system. The range of application of these methods is on the order of angstroms to nanometers. Examples of this type of work are the prediction of reaction pathways using electronic structure methods, the study of protein structure using molecular dynamics or Monte Carlo techniques, or the study of phase transitions in liquids and solids from knowledge of intermolecular forces.[2] At the mesoscopic level, coarse-grained models and mean field treatments are used to predict structure and properties at length scales ranging from tens of nanometers to microns. Examples of this type of research are the calculation of morphology in self-assembling systems (e.g., block copolymers and surfactants) and the study of macromolecular configuration (e.g., DNA) in microfluidic devices.[3,4,5] At the continuum or macroscopic level, one is interested in predicting the behavior of fluids and materials on laboratory scales (microns to centimeters), and this is usually achieved by numerical solution of the relevant conservation equations (e.g., Navier-Stokes, in the case of fluids).[6]

Over the last decade considerable progress has been achieved in the three categories described above. It is now possible to think about "multiscale modeling" approaches, in which distinct methods appropriate for different length scales are combined or applied simultaneously to achieve a comprehensive description of a system. *This progress has been partly due to the ever-increasing power of computers but, to a large extent, it has been the result of important theoretical and algorithmic developments in the area of computational materials and fluids modeling.*

Much of the interest in multiscale modeling methods is based on the premise that, one day, the behavior of entirely new materials or complex fluids will be

---

[1] De Pablo J. J.; Escobedo, F. A. *AIChE Journal* **2002**, *48*, 2716-2721.
[2] Greeley J.; Norskov, J. K.; Mavrikakis, M. *Ann. Rev. Phys. Chem.* **2002**, *53*, 319-348.
[3] Fredrickson G. H., Ganesan, V.; Drolet, F. *Macromol.* **2002**, *35*, 16-39.
[4] Hur J. S.; Shaqfeh, E. S. G.; Larson, R. A. *J. Rheol.* **2000**, *44*, 713-742.
[5] Jendrejack R. M.; de Pablo, J. J.; Graham, M. D. *J. Chem. Phys.* **2002**, *116*, 7752-7759.
[6] Bird, R. B.; Stewart, W. E.; Lightfoot, E. N. *Transport Phenomena:* 2nd Ed., John Wiley: New York, NY; 2002.

conceived or understood from knowledge of their atomic or molecular constituents. The goal of this brief document is to summarize how molecular structure and thermodynamic properties can be simulated numerically, to establish the promise of modern molecular simulation methods, including the opportunities they offer and the challenges they face, and to discuss how the resulting molecular-level information can be used in conjunction with mesoscopic and continuum modeling techniques for study of macroscopic systems.

As a first concrete example, it is instructive to consider the simulation of a phase diagram (a simple liquid-vapor coexistence curve) for a simple fluid (e.g., argon) from knowledge of the interactions between molecules; in the early 1990s, the calculation of a few points on such a diagram required several weeks of supercomputer time.[7] With more powerful techniques and faster computers, it is now possible to generate entire phase diagrams for relatively complex, industrially relevant models of fluids (such as mixtures of large hydrocarbons) in relatively short times (on the order of hours or days).[8]

A molecular simulation consists of the model or force field that is adopted to represent a system and the simulation technique that is employed to extract quantitative information (e.g., thermodynamic properties) about that model. For molecular simulations of the structure and thermodynamic properties of complex fluids and materials, particularly those consisting of large, articulated molecules (e.g., surfactants, polymers, proteins), Monte Carlo methods offer attractive features that make them particularly effective. In the context of molecular simulations, a Monte Carlo algorithm can be viewed as a process in which random steps or displacements (also known as "trial moves") away from an arbitrary initial state of the system of interest are carried out to generate large ensembles of realistic, representative configurations. There are two essential ingredients to any Monte Carlo algorithm: the first consists of the types of "moves" or steps that are used, and the second consists of the formalism or criteria that are used to guide an algorithm toward thermodynamic equilibrium.

The possibility of designing unphysical "trial moves," in which molecules can be temporarily destroyed and reassembled, often permits efficient exploration of the configuration space available to a system. In the case of long hydrocarbons, surfactants, or phospholipids, for example, configurational-bias techniques[9,10] can accelerate the convergence of a simulation algorithm by several orders of magnitude. In the case of polymers, re-bridging techniques permit simulation of structural properties that would not be accessible by any other means.[11]

---

[7]Panagiotopoulos, A. Z. *Mol.Sim.* **1992**, *9*, 1-23.
[8]Nath S.; Escobedo, F. A.; Patramai, I.; de Pablo, J. J. *Ind. Eng. Chem. Res.* **1998**, *37*, 3195.
[9]De Pablo, J. J.; Yan, Q.; Escobedo, F. A. *Ann. Rev. Phys. Chem.* **1999**, *50*, 377-411.
[10]Frenkel D.; Smit, B. *Understanding Molecular Simulation*, 2nd Ed., Elsevier Science: London, UK, 2002.
[11]Karayiannis N. C.; Mavrantzas, V. G.; Theodorou, D. N. *Phys., Rev. Lett.* **2002**, *88 (10)*, 105503.

The development of novel, clever Monte Carlo trial moves for specific systems is a fertile area of research; significant advances in our ability to model fluids and materials will result from such efforts.

A Monte Carlo simulation can be implemented in a wide variety of ways. In the most common implementation, trial moves are accepted according to probability criteria (the so-called Metropolis criteria) constructed in such a way as to result in an ensemble of configurations that satisfies the laws of physics. There is, however, considerable flexibility in the way in which such criteria are implemented. In expanded ensemble methods, for example, fictitious intermediate states of a system are created in order to facilitate transitions between an initial and a final state; transitions between states are accepted or rejected according to well-defined probability criteria. In parallel-tempering or replica-exchange simulation methods, calculations are conducted simultaneously on multiple replicas of a system. Each of these replicas can be studied at different conditions (e.g., different temperature); configuration exchanges between distinct replicas can be proposed and accepted according to probability criteria that guarantee that correct ensembles are generated for each replica. Within this formalism, a system or replica that is sluggish and difficult to study at low temperatures (e.g., a highly viscous liquid or a glassy solid), can benefit from information generated in high-temperature simulations, where relaxation and convergence to equilibrium are much more effective.[12] More recently, density-of-states techniques have been proposed as a possible means to generate all thermodynamic information about a system over a broad range of conditions from a single simulation.[13,14,15] *These examples serve to illustrate that if recent history is an indication of progress to come, new developments in the area of Monte Carlo methods will continue to increase our ability to apply first principles information to thermodynamic property and structure prediction, thereby supplementing and sometimes even replacing more costly experimental work. Such new developments will also facilitate considerably our ability to design novel and advanced materials and fluids from atomistic- and molecular-level information.*

As alluded to earlier, the results and predictions of a molecular simulation are only as good as the underlying model (or force field) that is adopted to represent a system. Transferable force fields are particularly attractive because, in the spirit of "group-contribution" approaches, they permit study of molecules and many-body systems of arbitrary complexity through assembly of atomic or chemical-group building blocks. Most force fields use electronic structure methods to generate an energy surface that is subsequently fitted using simple functional

---

[12]Yan, Q.; de Pablo, J. J.; *J. Chem. Phys.* **1999.** *111,* 9509.
[13]Wang, F. G.; Landau, D. P.; *Phys. Rev. Lett.* **2001,** *86 (10),* 2050-2053.
[14]Yan, Q. L.; Faller, R.; de Pablo, J. J. *J. Chem. Phys.* **2002,** *116,* 8745-8749.
[15]Yan, Q. L.; de Pablo, J. *Phys. Rev. Lett.* **2003,** *90 (3),* 035701.

forms. The resulting functions and their corresponding parameters are referred to as a force field. It is important to note that "raw" parameters extracted from electronic structure calculations are generally unable to provide a quantitative description of the structure and properties of complex fluids and materials; their subsequent optimization by analysis of experimental data considerably improves their applicability and predictive capability. Improvements in the efficiency of simulation techniques have rendered this last aspect of force field development much more tractable than it was a decade ago. *Reliable force fields are now being proposed for a wide variety of systems, including hydrocarbons, carbohydrates, alcohols, polymers, etc.*[6,16,17,18] *Accurate force fields are the cornerstone of fluids and materials modeling; much more work in this area is required to reach a stage at which modeling tools can be used with confidence to interpret the results of experiments and to anticipate the behavior of novel materials.*

The above discussion has been focused on Monte Carlo methods. Such methods can be highly effective for determining the equilibrium structure and properties of fluids and materials, but they do not provide information about time-dependent processes. Molecular dynamics (MD) methods, which are the technique of choice for study of dynamic processes, have also made considerable progress over the last decade. The development of multiple-time-step methods, for example, has increased significantly the computational efficiency of MD simulations. Unfortunately, however, the time scales amenable to molecular dynamics simulations are still on the order of tens or hundreds of nanoseconds. *Many of the processes of interest in chemistry and chemical engineering occur on much longer time scales (e.g., minutes or hours); it is unlikely that the several orders of magnitude that now separate our needs from what is possible with atomistic-level methods will be bridged by the availability of faster computers. It is therefore necessary to develop theoretical and computational methods to establish a systematic connection between atomistic and macroscopic time scales. These techniques are often referred to as multiscale methods or coarse-graining methods.*

While multiscale modeling is still in its infancy, its promise is such that considerable efforts should be devoted to its development in the years to come. A few examples have started to appear in the literature. In the case of solid materials, the challenge of coupling atomistic phenomena (e.g., the tip of a crack) with mechanical behavior (e.g., crack propagation and failure) over macroscopic domains has been addressed by several authors.[19,20] In the case of fluids, molecular-level structure (e.g., the conformation of DNA molecules in solution) has been

---

[16]Jorgensen, W. L.; Maxwell, D. S.; TiradoRives, J. *J. Amer. Chem. Soc.* **1996**, *118*, 11225-11236.
[17]Errington, J. R.; Panagiotopoulos, A. Z. *J. Phys. Chem B* **1999**, *103*, 6314-6322.
[18]Wick, C. D. Martin, M. G.; Siepmann, J. I. *J. Phys. Chem. B* **2000**, *104*, 8008-8016.
[19]Broughton, J. Q. Abraham, F. F.; N. Bernstein; *Phys. Rev. B* **1999**, *60*, 2391-2403.
[20]Smith, G. S., E. B. Tadmor, N. Bernstein; Kaxiras. E. *Acta Mater.* **2001**, *49*, 4089-4101.

solved concurrently with macroscopic flow problems (fluid flow through macro- and microfluidic geometries).[21,22] Multiscale methods for the study of dynamic processes currently rely on separation of time scales for various processes. One of the cornerstones of these methods is the averaging or coarse-graining of fast, local processes into a few, well-chosen variables carrying sufficient information content to provide a meaningful description of a system at longer time and length scales. The reverse is also true, and perhaps more challenging; information from a coarse-grained level must be brought back onto a microscopic level in a sensible manner, without introducing spurious behavior. This latter "reverse" problem is often underspecified and represents a considerable challenge. *New and better numerical schemes to transfer information between different description levels should be developed; this must be done without adding systematic perturbations to the system and in a computationally robust and efficient way. A better understanding of coarse graining techniques is sorely needed, as are better algorithms to merge different levels of description in a seamless and correct manner.*

---

[21]Jendrejack, RM, de Pablo, J. J.; Graham, M. D. *J. Non-Newton Fluid* **2002**, *108*, 123-142.
[22]Jendrejack, R. M.; Graham. M. D.; de Pablo, J. J. Multiscale simulation of DNA solutions in microfluidic devices, unpublished.

ADVANCES IN INFORMATION & COMMUNICATION TECHNOLOGIES: OPPORTUNITIES AND CHALLENGES IN CHEMICAL SCIENCE AND TECHNOLOGY

Thom H. Dunning, Jr.
University of Tennessee and Oak Ridge National Laboratory

Introduction

The topics to be covered in this paper are the opportunities in chemical science and technology that are arising from the advances being made in computing technologies—computers, data stores, and networks. That computing technology continues to advance at a dizzying pace is familiar to all of us. Almost as soon as one buys a PC, it becomes outdated because a faster version of the microprocessor that powers that PC becomes available. Likewise, tremendous advances are being made in memory and data storage capacities. The density of memory is increasing at the same rate as computing speed, and disk drive densities are increasing at an even faster pace. These advances have already led to a new era in computational chemistry—computational models of molecules and molecular processes are now so widely accepted and PCs and workstations so reasonably priced that molecular calculations are routinely used by many experimental chemists to better understand the results of their experiments. For someone like the author, who started graduate school in 1965, this transformation of the role of computing in chemistry is little short of miraculous.

Dramatic increases are also occurring in network bandwidth. Bandwidth that was available only yesterday to connect computers in a computer room is becoming available today between distant research and educational institutions—the National Light Rail is intended to link the nation's most research-intensive universities at 10 gigabits per second. This connectivity has significant implications for experimental and computational chemistry. Collaborations will grow as geographically remote collaborators are able to easily share data and whiteboards, simultaneously view the output of calculations or experiments, and converse with one another face to face through audio-video links. Data repositories provided by institutions such as the National Institute of Standards and Technology, although not nearly as large in chemistry as in molecular biology, are nonetheless important and will become as accessible as the data on a local hard drive. Finally, network advances promise to make the remote use of instruments, especially expensive one-of-a-kind or first-of-a-kind instruments, routine, enabling scientific advances across all of chemical science.

At the current pace of change, an *order-of-magnitude increase in computing and communications capability* will occur *every five years*. An order-of-magnitude increase in the performance of any technology is considered to be the threshold for revolutionary changes in usage patterns. It is important to keep this in mind. In fact, Professor Jack Dongarra of the University of Tennessee, one of the world's foremost experts in scientific computing, has recently stated this fact in stark terms:

> The rising tide of change [resulting from advances in information technology] shows no respect for the established order. Those who are unwilling or unable to adapt in response to this profound movement not only lose access to the opportunities that the information technology revolution is creating, they risk being rendered obsolete by smarter, more agile, or more daring competitors.

Never before in the history of technology have revolutionary advances occurred at the pace being seen in computing. It is critical that chemical science and technology in the United States aggressively pursue the opportunities offered by the advances in information and communications technologies. Only by doing so will it be able to maintain its position as a world leader.

It is impossible to cover all of the opportunities (*and challenges*) in chemical science and technology that will result from the advances being made in information and communications technologies in a 30-minute presentation. Instead, I focus on three examples that cover the breadth of opportunities that are becoming available: (*i*) advances in computational modeling that are being driven by advances in computer and disk storage technology, (*ii*) management of large datasets that is being enabled by advances in storage and networking technologies, and (*iii*) remote use of one-of-a-kind or first-of-a-kind scientific instruments resulting from advances in networking and computer technologies. In this paper, I focus on applications of high-end computing and communications technologies. However, it is important to understand that the "trickle-down effect" is very real in computing (if not in economics)—the high-end applications discussed today will be applicable to the personal computers, workstations, and departmental servers of tomorrow.

Computational Modeling in Chemical Science and Technology

First, consider the advances in computational modeling wrought by the advances in computer technology. Almost everybody is aware of Moore's Law, the fact that every 18-24 months there is a factor of 2 increase in the speed of the microprocessors that are used to drive personal computers as well as many of today's supercomputers. Within the U.S. Department of Energy, the increase in computing power available for scientific and engineering calculations is codified in the "ASCI Curve," which is a result of Moore's law compounded by the use of increased number of processors (Figure 1). The result is computing power that is well beyond that predicted by Moore's law alone. If there is a factor of 2 increase every 18-24 months from computer technology and the algorithms used in scientific and engineering calculations scale to twice as many processors over that same period of time, the net result is a factor of 4 increase in computing capabil-

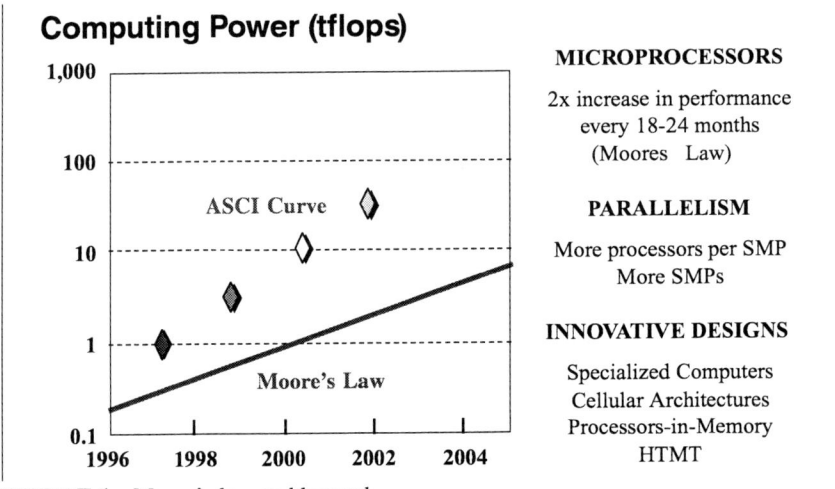

FIGURE 1 Moore's law and beyond.

ity. At this pace, one can realize an order-of magnitude-increase in computing power in just five years.

The increase in computing power in the past five years, in fact, follows this pattern. In 1997, "ASCI Red" at Sandia National Laboratories was the first computer system capable of performing more than 1 trillion arithmetic operations per second (1 teraflops). ASCI Red had a total of 9216 Pentium Pro processors, a peak performance of 1.8 teraflops, and 0.6 terabytes of memory. In 2002, the Japanese Earth Simulator is the world's most powerful computer with 5120 processors, a peak performance of 40.96 teraflops, and more than 10 terabytes of memory (Figure 2). This is an increase of a factor of over 20 in peak performance in just five years! But, the increase in delivered computing power is even greater. On ASCI Red a typical scientific calculation achieved 10-20% of peak performance, or 100-200 gigaflops. On the Japanese Earth Simulator, it is possible to obtain 40-50% of peak performance, or 16 to 20 teraflops. Thus, in terms of *delivered performance*, the increase from 1997 to 2002 is more like a factor of 100.

The Earth Simulator is the first supercomputer in a decade that was designed for science and engineering computations. Most of the supercomputers in use today were designed for commercial applications, and commercial applications place very different demands on a computer's processor, memory, I/O, and interconnect subsystems than science and engineering applications. The impact of this is clearly evident in the performance of the Earth Simulator. For example, the performance of the Earth Simulator on the LinPack benchmark[1] is 35.86 teraflops

---

[1] See: *http://www.top500.org/list/2002/11/*.

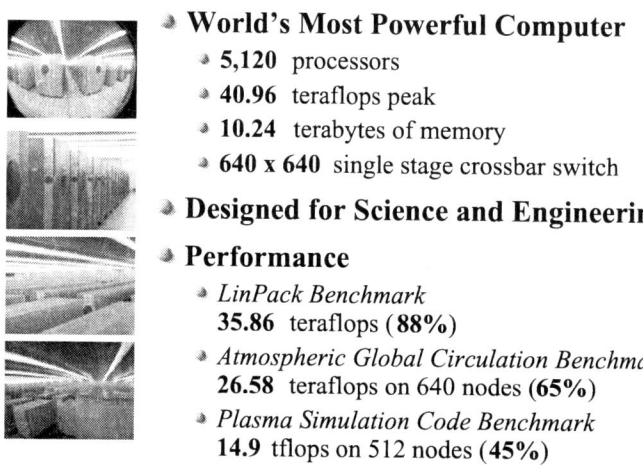

- **World's Most Powerful Computer**
  - 5,120 processors
  - 40.96 teraflops peak
  - 10.24 terabytes of memory
  - 640 x 640 single stage crossbar switch
- **Designed for Science and Engineering**
- **Performance**
  - *LinPack Benchmark*
    35.86 teraflops (**88%**)
  - *Atmospheric Global Circulation Benchmark*
    26.58 teraflops on 640 nodes (**65%**)
  - *Plasma Simulation Code Benchmark*
    14.9 tflops on 512 nodes (**45%**)

FIGURE 2   Japanese Earth Simulator.

or 87.5% of peak performance (Figure 3). The Earth Simulator is a factor of 5 faster on the LinPack Benchmark than its nearest competitor, ASCI White, even though the difference in peak performance is only a factor of three. The performance of the Earth Simulator on the LinPack benchmark exceeds the integrated total of the three largest machines in the United States (ASCI White, LeMieux at the Pittsburgh Supercomputing Center, and NERSC3 at Lawrence Berkeley National Laboratory) by a factor of nearly 2.5.

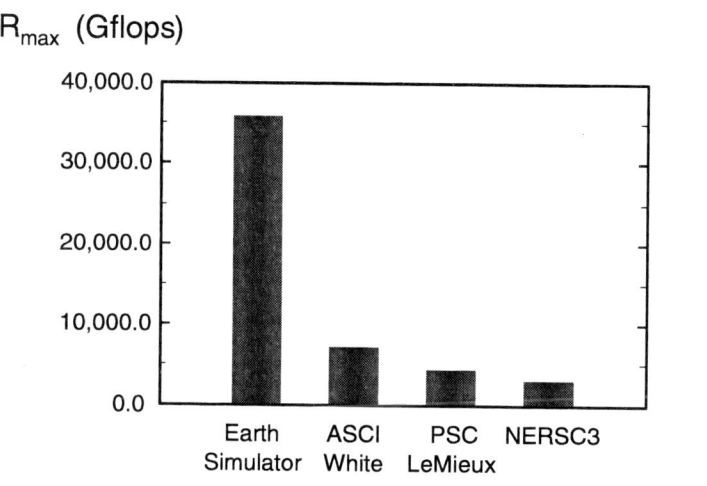

FIGURE 3   LinPack benchmarks.

The performance of the Earth Simulator is equally impressive when real scientific and engineering applications are considered. A general atmospheric global circulation benchmark ran at over 26 teraflops, or 65% of peak performance, on the Earth Simulator.[2] On the commercially oriented machines that are currently being used in the United States, it has proven difficult to achieve more than 10% of peak performance for such an application.[3] So, not only is the raw speed of the Japanese Earth Simulator very high, it is also very effective for scientific and engineering applications. The Earth Simulator is expected to run a wide range of scientific and engineering applications 10-100 times faster than the fastest machines available in the United States. As Professor Dongarra noted, drawing a comparison with the launching of *Sputnik* by the Soviet Union in 1958, the Earth Simulator is the "computernik" of 2002, representing a wake-up call illustrating how computational science has been compromised by supercomputers built for commercial applications.

The above discussion has focused on general computer systems built using commodity components. Computer companies now have the capability to design and build specialized computers for certain types of calculations (e.g., molecular dynamics) that are far more cost-effective than traditional supercomputers. Examples of these computers include MD-GRAPE[4] for molecular dynamics calculations and QCDOC[5] for lattice gauge QCD calculations. Price-performance improvements of orders of magnitude can be realized with these specialized computers, although with some loss of flexibility and generality. However, with the exception of the lattice gauge QCD community, specialized computers have not gained widespread acceptance in science and engineering. This is due largely to the rapid increases in the computing capabilities of general-purpose microprocessors over the last couple of decades—with an increase of a factor of 2 every 18-24 months, the advantages offered by specialized computers can rapidly become outdated. In addition, the limited flexibility of specialized computers often prevented the use of new, more powerful algorithms or computational approaches. However, given the increasing design and fabrication capabilities of the computer industry, I believe that this is a topic well worth reexamining.

---

[2]S. Shingu, H. Takahara, H. Fuchigami, M. Yamada, Y. Tsuda, W. Ohfuchi, Y. Sasaki, K. Kobayashi, T. Hagiwara, S. Habata, M. Yokokawa, H. Itoh, and K. Otsuka, "A 26.58 Tflops Global Atmospheric Simulation with the Spectral Transform Method on the Earth Simulator," presented at SC2002 (Nov. 2002).

[3]"Capacity of U.S. Climate Modeling to Support Climate Change Assessment Activities," Climate Research Committee, National Research Council, (National Academy Press, Washington, 1999).

[4]See: http://www.research.ibm.com/grape/.

[5]D. Chen, N. H. Christ, C. Cristian, Z. Dong, A. Gara, K. Garg, B. Joo, C. Kim, L. Levkova, X. Liao, R. D. Mawhinney, S. Ohta, and T. Wettig, *Nucl. Phys. B (Proc. Suppl.)* **2001**, *94*, 825-832; P. A. Boyle, D. Chen, N. H. Christ, C. Cristian, Z. Dong, A. Gara, B. Joo, C. Kim, L. Levkova, X. Liao, G. Liu, R. D. Mawhinney, S. Ohta, T. Wettig, and A. Yamaguchi, *Nucl. Phys. B (Proc. Suppl.)* **2002**, *106*, 177-183.

In summary, it is clear that over the next decade and likely longer, we will see continuing increases in computing power, both from commercially oriented computers and from more specialized computers, including follow-ons to the Earth Simulator. With the challenge offered by the Japanese machine, U.S. computer companies can be expected to refine their designs in order to make their computers more appropriate for scientific and engineering calculations, although a major change would be required in the business plans of companies such as IBM and HP to fully respond to this challenge (the scientific and engineering market is miniscule compared to the commercial market). These increases in computing power will have a major impact on computer-based approaches to science and engineering.

Opportunities in Computational Chemistry

What do the advances occurring in computing technology mean for computational chemistry, especially electronic structure theory, the author's area of expertise? It is still too early to answer this question in detail, but the advances in computing technology will clearly have a major impact on both the size of molecules that can be computationally modeled and the fidelity with which they can be modeled. Great strides have already been achieved for the latter, although as yet just for small molecules (the size of benzene or smaller). Consider, for example, the calculation of bond energies (Figure 4). When I started graduate school

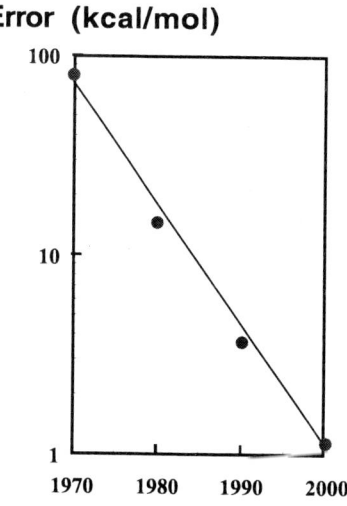

- Bond energies critical for describing many chemical phenomena
- Accuracy of calculated bond energies increased dramatically from 1970 - 2000
- Due to advances in
  - Theoretical methodology
  - Computational techniques
  - Computing technology

FIGURE 4 Opportunity: increased fidelity of molecular models.

in 1965, we were lucky to be able to calculate bond energies accurate to 50 kcal/mol. Such predictions were not useful for much of anything and, in fact, looking back on the situation, I am amazed that experimental chemists showed such tolerance when we reported the results of such calculations. But the accuracy of the calculations has been improving steadily over the past 30 years. By 2000, bond energies could be predicted to better than 1 kcal/mol, as good as or better than the accuracy of most experimental measurements.

This increase in the accuracy of calculated bond energies is not just due to advances in computing technology. Computing advances were certainly important, but this level of accuracy could never have been achieved without significant advances in theoretical methodology (e.g., the development of coupled cluster theory) as well as in computational techniques, for example, the development of a family of basis sets that systematically converge to the complete basis set (CBS) limit. Coupled-cluster theory, which was first introduced into chemistry in 1966 (from nuclear theory) by J. Cizek and subsequently developed and explored by J. Paldus, I. Shavitt, R. Bartlett, J. Pople, and their collaborators, provides a theoretically sound, rapidly convergent expansion of the wave function of atoms and molecules.[6] In 1989, K. Ragavachari, J. Pople and co-workers suggested a perturbative correction to the CCSD method to account for the effect of triple excitations.[7] The accuracy of the CCSD(T) method is astounding. Dunning[8] has shown that it provides a nearly quantitative description of molecular binding from such weakly bound molecules as He_2, which is bound by only 0.02 kcal/mol, to such strongly bound molecules as CO, which is bound by nearly 260 kcal/mol—a range of four orders of magnitude. One of the most interesting aspects of coupled cluster theory is that it is the mathematical incarnation of the electron pair description of molecular structure—a model used by chemists since the early twentieth century. True to the chemist's model, the higher-order corrections in coupled cluster theory (triple and higher excitations) are small, although not insignificant.

The accuracy of the CCSD(T) method for strongly bound molecules is illustrated in Figure 5. This figure provides a statistical analysis of the errors in the computed D_e values for a representative group of molecules.[9] The curves represent the normal error distributions for three different methods commonly used to solve the electronic Schrödinger equation: second-order Møller-Plesset perturbation theory (MP2), coupled cluster theory with single and double excitations, and

---

[6]A brief history of coupled cluster theory can be found in R. J. Bartlett, *Theor. Chem. Acc.* **2000**, *103*, 273-275.

[7]K. Raghavachari, G. W. Trucks, J. A. Pople, and M. Head-Gordon, *Chem. Phys. Lett.* **1989**, *157*, 479-483.

[8]T. H. Dunning, Jr., *J. Phys. Chem. A* **2000**, *104*, 9062-9080.

[9]K. L. Bak, P. Jørgensen, J. Olsen, T. Helgaker, and W. Klopper, *J. Chem. Phys.* **2000**, *112*, 9229-9242.

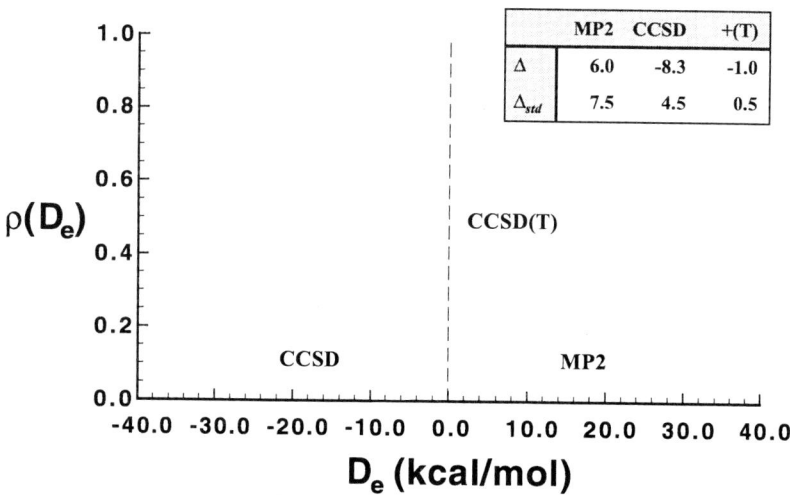

FIGURE 5 Theoretical advances: coupled cluster calculation of D_e values.

CCSD with the perturbative correction for triple excitations, CCSD(T). The box at the top right lists the average error (Δ_{ave}) and the standard deviation (Δ_{std}). If the Schrödinger equation was being solved exactly for this group of molecules, the standard error distribution would simply be a δ-function at zero (0.0). It is clear that the error distributions for the MP2 and CCSD methods do not have this shape. Not only are the average errors far from zero (Δ_{ave} = 6.0 and –8.3 kcal/mol, respectively), but the widths of the error distributions are quite large (Δ_{std} = 7.5 and 4.5 kcal/mol, respectively). Clearly, neither the MP2 nor the CCSD method is able to provide a consistent description of the molecules included in the test set, although the CCSD method, with its smaller Δ_{std}, can be considered more reliable than the MP2 method. On the other hand, if the perturbative triples correction is added to the CCSD energies, the error distribution shifts toward zero (Δ_{ave} = –1.0 kcal/mol) and sharpens substantially (Δ_{std} = 0.5 kcal/mol). The accuracy of the CCSD(T) method is not a fluke, as studies with the CCSDT and CCSDTQ methods, limited though they may be, show.

The above results illustrate the advances in our ability to quantitatively describe the electronic structure of molecules. This advance is in large part due to our ability to converge the solutions of the electronic Schrödinger equation. In the past, a combination of incomplete basis sets and lack of computing power prevented us from pushing calculations to the complete basis set (CBS) limit. For any finite basis set calculation, the error is the sum of the error due to the use of an approximate electronic structure method plus the error due to the use of an incomplete basis set. These errors can be of opposite sign, which can lead to confusing "false positives" (i.e., agreement with experiment due to cancellation of

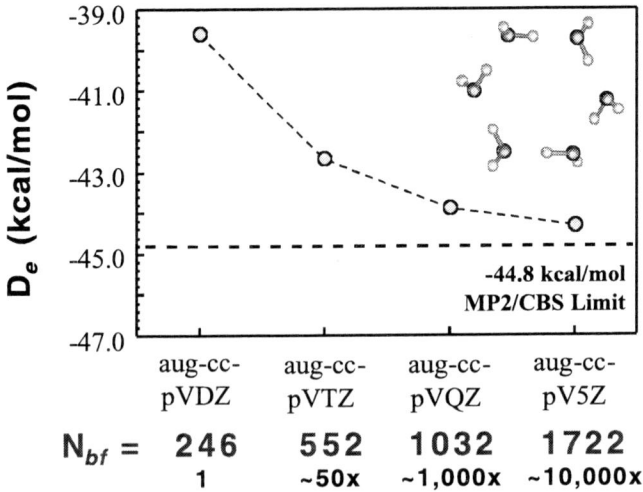

FIGURE 6 Opportunity: converged molecular calculations (data from Xantheas S.; Burnham, C.; Harrison, R. *J. Chem. Phys.* **2002**, *116*, 1493).

errors).[8] At the CBS limit on the other hand, the remaining error is that due to the method itself. Thus, the ability to push calculations to the CBS limit has allowed us to separate the methodological errors from the basis set truncation errors, greatly advancing our understanding of the accuracy of the various approximate methods used to solve the Schrödinger equation.

The computing resources needed to push calculations to the CBS limit are illustrated in Figure 6. This figure displays the results of MP2 calculations[10] on the water hexamer by Xantheas, Burnham, and Harrison.[11] In addition to its intrinsic importance, an accurate value of the binding energy for the water hexamer was needed to provide information for the parameterization of a new many-body potential for water.[12] This information is not available from experiment and thus its only source was high-level quantum chemical calculations. The energy plotted in the figure is that required to dissociate the water hexamer at its equilibrium geometry into six water molecules at their equilibrium geometries: $E_e[(H_2O)_6]$—$6Ee(H_2O)$. With the augmented double zeta basis set (aug-cc-pVDZ), the calculations predict that the hexamer is bound by 39.6 kcal/mol. Calculations with the

---

[10] In contrast to chemically-bound molecules, the MP2 method predicts binding energies for hydrogen-bonded molecules such as $(H_2O)n$ comparable to those obtained with the CCSD(T) method (see Ref. 8).

[11] S. S. Xantheas, C. J. Burnham, and R. J. Harrison, *J. Chem. Phys.* **2002**, *116*, 1493-1499.

[12] C. J. Burnham and S. S. Xantheas, *J. Chem. Phys.* **2002**, *116*, 1500-1510; C. J. Burnham and S. S. Xantheas, *J. Chem. Phys.* **2002**, *116*, 5115-5124.

quintuple zeta basis set (aug-cc-pV5Z), on the other hand, predict an equilibrium binding energy of 44.3 kcal/mol, nearly 5 kcal/mol higher. As can be seen, the variation of the binding energy with basis set is so smooth that the value can be extrapolated to the CBS limit. This yields a binding energy of 44.8 kcal/mol. From benchmark calculations on smaller water clusters, this number is expected to be accurate to better than 1 kcal/mol.

Pushing molecular calculations on molecules such as $(H_2O)_6$ to the CBS limit requires substantial computing resources. If the amount of computer time required for a calculation with the double zeta set is assigned a value of 1 unit, then a calculation with the triple zeta set requires 50 units. The amount of computing time required escalates to the order of 1000 units for the quadruple zeta set and the order of 10,000 units for the quintuple zeta set. Clearly, advances in computing power are contributing significantly to our ability to provide quantitative predictions of molecular properties.

Another opportunity provided by greatly increased computing resources is the ability to extend calculations such as those described above to much larger molecules. This is important in the interpretation of experimental data (experimentalists always seem to focus on larger molecules than we can model), to characterize molecules for which experimental data is unavailable, and to obtain data for parameterizing semiempirical models of more complex systems. The latter was the driving force in the calculations on $(H_2O)n$ referred to above.[11] This was also what drove the study of water interacting with cluster representations of graphite sheets by Feller and Jordan[13] (Figure 7). The MP2 calculations reported by these authors considered basis sets up to aug-cc-pVQZ and graphitic clusters up to $(C_{96}H_{24})$. By exploiting the convergence properties of the correlation consistent basis sets as well as that of the graphitic clusters, they predicted the equilibrium binding energy for water to graphite to be 5.8 ± 0.4 kcal/mol. The largest MP2 calculation reported in this study was on H_2O-$C_{96}H_{24}$ and required 29 hours on a 256-processor partition of the IBM SP2 at PNNL's Environmental Molecular Sciences Laboratory.

The ultimate goal of Feller and Jordan's work is to produce a potential that accurately represents the interaction of water with a carbon nanotube. Their next study will involve H_2O interacting with a cluster model of a carbon nanotube. However, this will require access to the next generation massively parallel computing system from Hewlett-Packard that is currently being installed in EMSL's Molecular Science Computing Facility.[14]

---

[13] D. Feller and K. Jordan, *J. Phys. Chem. A* **2000**, *104*, 9971-9975.
[14] See: http://www.emsl.pnl.gov:2080/capabs/mscf_capabs.html and http://www.emsl.pnl.gov:2080/capabs/mscf/hardware/config_hpcs2.html.

H_2O-$C_{96}H_{24}$

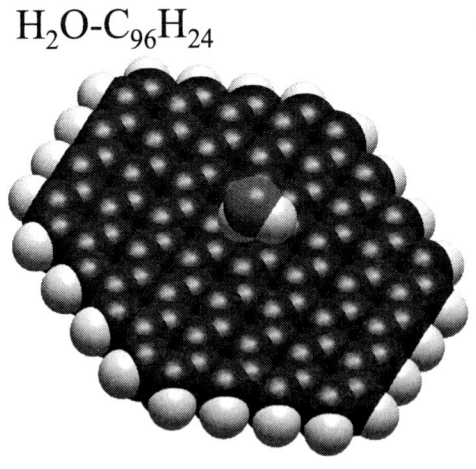

D_e = -5 kcal/mol

▪ Need interaction potentials to model nanoscale processes

▪ Little data from experiment, need accurate calculations

▪ First step: H_2O-$C_{96}H_{24}$; next step H_2O with segment of nanotube.

FIGURE 7 Opportunity: Larger, More Complex Molecules. Courtesy of D. Feller, Pacific Northwest National Laboratory; see also Feller, D.; Jordan, K. D. *J. Phys. Chem. A* **2000**, *104*, 9971.

Challenges in Computational Chemistry

Although the opportunities offered by advances in computing technologies are great, many challenges must be overcome to realize the full benefits of these advances. Most often the focus is on the *computational challenges* because these are indeed formidable. However, we must also quantify the limitations of the existing theories used to describe molecular properties and processes as well as develop new theories for those molecular phenomena that we cannot adequately describe at present (*theoretical challenges*). In addition, we must seek out and carefully evaluate new mathematical approaches that show promise of reducing the cost of molecular calculations (*mathematical challenges*). The computational cost of current electronic structure calculations increases dramatically with the size of the molecule (e.g., the CCSD(T) method scales as N^7, where N is the number of atoms in the molecule). Of particular interest are mathematical techniques that promise to reduce the scaling of molecular calculations. Work is currently under way in all of these challenge areas.

As noted above, the theoretical challenges to be overcome include the validation of existing theories as well as the development of new theories. One of the surprises in electronic structure theory in the 1990s was the finding that Møller-Plesset perturbation theory, the most widely used means to include electron correlation effects, does not lead to a convergent perturbation expansion series. This

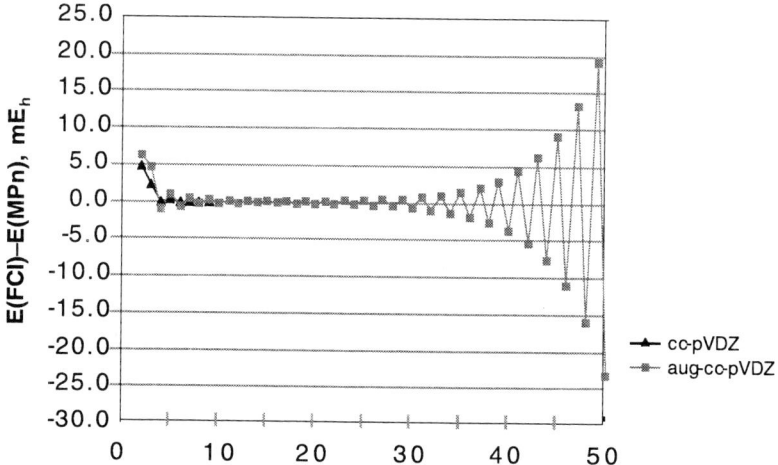

FIGURE 8 Theoretical challenges: convergence of perturbation expansion for HF (data from Olsen, J.; Christiansen, O.; Koch, H.; Jørgensen, P. *J. Chem. Phys.* **1996**, *105*, 5082-5090).

nature of the perturbation expansion is illustrated in Figure 8, which plots the difference between full CI and perturbation theory energies for the HF molecule in a double zeta and augmented double zeta basis set as a function of the order of perturbation theory. As can be seen, at first the perturbation series appears to be converging, but then, around fifth- or sixth-order perturbation theory, the series begins to oscillate (most evident for the aug-cc-pVDZ set) with the oscillations becoming larger and larger with increasing order of perturbation theory.[15] The perturbation series is clearly diverging even for hydrogen fluoride, a molecule well described by a Hartree-Fock wave function. Dunning and Peterson[16] showed that even at the complete basis set limit, the MP2 method often provides more accurate results than the far more computationally demanding MP4 method. Thus, the series is not well behaved even at low orders of perturbation theory. How many other such surprises await us in theoretical chemistry?

Although we now know how to properly describe the ground states of molecules,[17] the same cannot be said of molecular excited states. We still await an

---

[15] J. Olsen, O. Christiansen, H. Koch, and P. Jørgensen, *J. Chem. Phys.* **1996**, *105*, 5082-5090.

[16] T. H. Dunning, Jr. and K. A. Peterson, J. Chem. Phys. 108, 4761-4771 (1998).

[17] This is not to say that the CCSD(T) method provides an accurate description of the ground states of *all* molecules. The coupled cluster method is based on a Hartree-Fock reference wave function and thus will fail when the HF wave function does not provide an adequate zero-order description of the molecule. The development of multireference coupled cluster methods is being actively pursued by several groups.

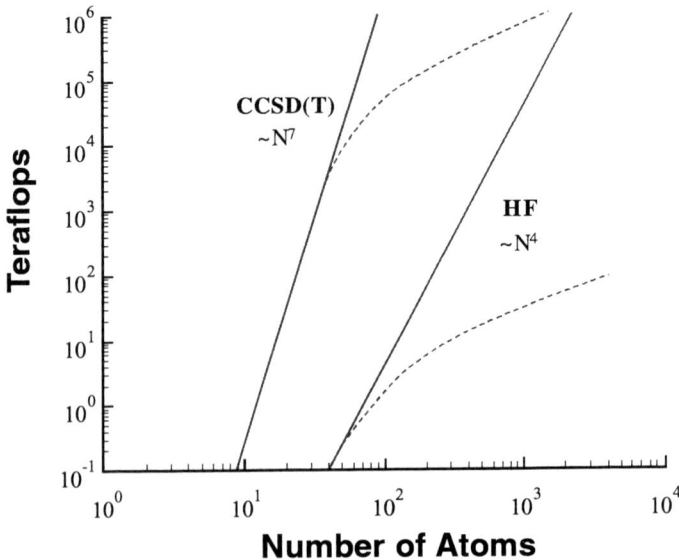

FIGURE 9 Mathematical challenges: scaling laws for molecular calculations.

accurate, general, computationally tractable approach for solving for the higher roots of the electronic Schrödinger equation. Without such a theory, many molecular phenomena, such as visible-UV spectroscopy and photochemistry, will remain out of modeling reach.

As noted above, current molecular electronic structure calculations scale as a high power of the number of atoms in the molecule. This is illustrated in Figure 9. For example, the Hartree-Fock method scales as N^4, while the far more accurate CCSD(T) method scales as N^7. Thus, when using the CCSD(T) method, doubling the size of a molecule increases the computing time needed for the calculation by over two orders of magnitude. Even if computing capability is doubling every 18 to 24 months, it would take another 10 years before computers would be able to model a molecule twice as big as those possible today. Fortunately, we know that, for sufficiently large molecules, these methods don't scale as N^4 or N^7. For example, it has long been known that simple, controllable mathematical approximations such as integral screening can reduce the scaling of the HF method to N^2 for sufficiently large molecules. The impact of this is illustrated by the dashed curve in the figure.

Using more advanced mathematical techniques, such as the fast multipole method (FMM) to handle the long-range Coulomb interaction,[18] and a separate

---

[18] L. Greengard and V. Rohklin, *Acta Numerica* **6** (Cambridge University Press, Cambridge, 1997), pp 229-269 and references therein.

treatment of the exchange interaction, it is possible to develop a HF algorithm that exhibits linear scaling in the very large molecule limit.[19] Combining these techniques with Fourier methods provides improved scaling even for small systems in high-quality basis sets. These techniques can be straightforwardly extended to DFT calculations and similar reductions in the scaling laws are possible for correlated calculations, including CCSD(T) calculations. As in the HF method, it is possible to exploit screening in calculating the correlation energy. However, to take advantage of screening, the equations for the correlated methods must be rewritten in terms of the atomic orbital basis rather than the molecular orbital basis. This has recently been done for the coupled cluster method by Scuseria and Ayala,[20] who showed that, with screening alone, the CCD equations could be solved more efficiently in the AO basis than in the MO basis *for sufficiently large molecules*. Since the effectiveness of screening and multipole summation techniques increase with molecule size, the question is not whether the use of AOs will reduce the scaling of coupled cluster calculations but, rather, how rapidly the scaling exponent will decrease with increasing molecular size.

The impact of reduced scaling algorithms is just beginning to be felt in chemistry, primary in HF and DFT calculations.[21] However, reduced scaling algorithms for correlated molecular calculations are likely to become available in the next few years. When these algorithmic advances are combined with advances in computing technology, the impact will be truly revolutionary. Problems that currently seem intractable will not only become doable, they will become routine.

Numerical techniques for solving the electronic Schrödinger equation are also being pursued. Another paper from this workshop has been written by R. Friesner, who developed the pseudospectral method, an ingenious half numerical-half basis set method. Another approach that is being actively pursued is the use of wavelet techniques to solve the Schrödinger equation. R. Harrison and coworkers recently reported DFT-LDA calculations on benzene[22] that are the most accurate available to date (Figure 10). Unlike the traditional basis set expansion approach, the cost of wavelet-based calculations automatically scales linearly in the number of atoms. Although much remains to be done to optimize the codes for solving the Hartree-Fock and DFT equations, not to mention the development of wavelet approaches for including electron correlation effects via methods such as CCSD(T), this approach shows great promise for extending rigorous electronic structure calculations to large molecules.

Finally, there are computational challenges that must be overcome. The supercomputers in use today achieve their power by using thousands of proces-

---

[19]C. Ochsenfeld, C. A. White, and M. Head-Gordon, *J. Chem. Phys.* **1998**, *109*, 1663-1669.
[20]G. E. Scuseria and P. Y. Ayala, *J. Chem. Phys.* **1999**, *111*, 8330-8343.
[21]See, *e.g.*, the Feature Article by G. E. Scuseria, *J. Phys. Chem. A* **1999**, *103*, 4782-4790.
[22]R. J. Harrison, G. I. Fann, T. Yanai, and G. Beylkin, "Multiresolution Quantum Chemistry: Basic Theory and Initial Applications," *to be published*.

DFT/LDA Energy

10^{-3} -230.194
10^{-5} -230.19838
10^{-7} -230.198345

For comparison

Partridge-3 primitive set +
aug-cc-pVTZ polarization set:
-230.158382 hartrees

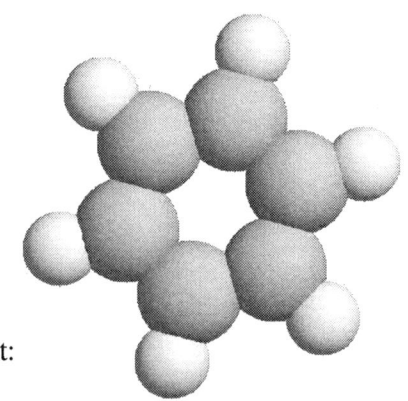

FIGURE 10 Mathematical challenges: multiwavelet calculations on benzene. Courtesy of R. Harrison, G. Fann, and G. Beylkin, Pacific Northwest National Laboratory.

sors, and computers are already in the design stage that use tens of thousands to hundreds of thousands of processors. So, one challenge is to develop algorithms that scale well from a few processors to tens of processors, to hundreds of processors, to thousands of processors, and so on. This is a nontrivial challenge and is an area of active research in computational science, computer science, and applied mathematics. In some cases, this work has been very successful with calculations using more than 1,000 processors being reported with NWChem[23] and NAMD.[24] For other scientific applications, we don't know when such algorithms will be discovered—this is, after all, an issue of creativity and creativity respects no schedules.

One of the surprises that I had as I began to delve into the problems associated with the use of the current generation of supercomputers was the relatively poor performance of many of the standard scientific algorithms on a single processor. Many algorithms achieved 10% or less of peak performance! This efficiency then further eroded as the calculations were scaled to more and more processors. Clearly, the cache-based microprocessors and their memory subsystems used in today's supercomputers are very different than those used in the supercomputers of the past. Figure 11 illustrates the problem. In the "Good Ole Days" (just a decade ago), Cray supercomputers provided very fast data paths between the central processing unit and main memory. On those machines, many vector

---

[23]See benchmark results at the NWChem web site *http://www.emsl.pnl.gov:2080/docs/nwchem/nwchem.html*.

[24]J. C. Phillips, G. Zheng, S. Kumar, and L. V. Kale, "NAMD: Biomolecular Simulation on Thousands of Processors," presented at SC2002 (Nov. 2002). See also *http://www.ks.uiuc.edu/Research/namd/*.

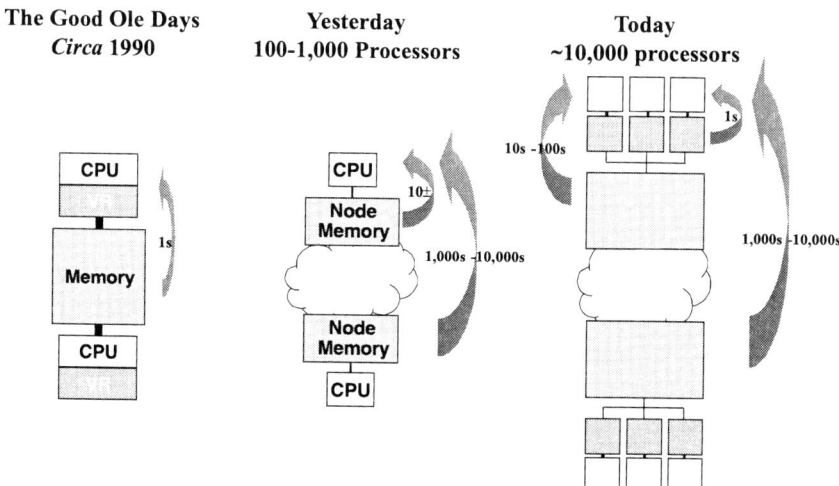

FIGURE 11 Computational challenges: keeping the processors busy. The numbers superimposed on the arrows refer to the number of processor cycles required to transfer a byte of data from the indicated memory location to the processor.

arithmetic operations, such as a DAXPY, which requires two words to be drawn from memory plus one written to memory every processor cycle, could be run at full speed directly from main memory. The supercomputers of today, on the other hand, are built from commercially oriented microprocessors and memory. Although this greatly reduces the cost of the supercomputers, the price is slow communications between the microprocessor and memory. Thus, it may take tens of cycles to transfer a word from memory to the processor. Computer designers attempt to minimize the impact of slow access to main memory by placing a fast cache memory between the processor and main memory. This works well if the algorithm can make effective use of cache, but many scientific and engineering algorithms do not. New cache-friendly algorithms must be developed to take full advantage of the new generation of supercomputers—again, a problem in creativity.

One of the reasons for the success of the Japanese Earth Simulator is that it was designed with a knowledge of the memory usage patterns of scientific and engineering applications. Although the bandwidth between the processor and memory in the Earth Simulator does not match that in the Cray computers of the 1990s (on a relative basis), it is much larger (four times or more) than that for LLNL's ASCI White or PSC's LeMieux. In late 2002, Cray, Inc. announced its new computer, the Cray X1. The Cray X1 has the same bandwidth (per flops) as the Earth Simulator, but each processor also has a 2 Mbyte cache with a bandwidth that is twice the bandwidth to main memory. The scientific and engineering community is eagerly awaiting the delivery and evaluation of this new computer from Cray, Inc. Oak Ridge National Laboratory is leading this effort for DOE's Office of Science.

The figure also illustrates the speed with which data can be transferred from the memory associated with other processors. This can require thousands or tens of thousands of processor cycles. The trick to developing scalable algorithms is to keep the data close to the processor(s) that need it. This is, of course, easier said than done.

Computational Modeling of Complex Chemical Systems

I don't know how many in the audience read John Horgan's book *The End of Science: Facing the Limits of Knowledge in the Twilight of the Scientific Age* (Little Brown & Company, 1997). Given the amazing advances in scientific knowledge that are being made each day, I began reading this book in a state of disbelief. It wasn't until I was well into the book that I realized that Horgan was not talking about the end of science, but rather the end of reductionism in science. These are not the same thing. A physicist might be satisfied that he understands chemistry once the Schrödinger equation had been discovered, but for chemists the job has only begun. Chemists are interested in uncovering how the basic laws of physics become the laws that govern molecular structure, energetics, and reactivity. Molecules are complex systems whose behavior, although a result of the fundamental laws of physics, cannot be directly connected to them.

Although I don't think we have yet reached the end of reductionism (much still remains to be discovered in physics, chemistry, and biology), I do think that we are in a period of transition. Scientists spent most of the twentieth century trying to understand the fundamental building blocks and processes that underlie our material world. The result has been a revolution—chemists now understand much about the basic interactions between atoms and molecules that influence chemical reactivity and are using that knowledge to create molecules that once could be obtained only from nature (e.g., cancer-fighting taxol); biologists now understand that the basic unit of human heredity is a large, somewhat monotonous molecule and have nearly determined the sequence of A's, T's, G's and C's in human DNA and are using this knowledge to pinpoint the genetic basis of diseases. As enlightening as this has been, however, we are now faced with another, even greater, challenge—assembling all of the information available on building blocks and processes in a way that will allow us to predict the behavior of complex, real-world systems. This can only be done by employing computational models. This is the scientific frontier of the twenty-first century. *We are not at the end of science, we are at a new beginning.* Although Horgan may not recognize this activity as science, the accepted definition of science, "knowledge or a system of knowledge covering general truths or the operation of general laws especially as obtained and tested through the scientific method," indicates that this is science nonetheless.

One example of a complex chemical system is an internal combustion engine. To predict the behavior of such a system, we must be able to model a wide variety of physical and chemical processes:

- simulate the mechanical device itself, including the dynamical processes involved in the operation of the device as well as the changes that will occur in the device as a result of combustion (changes in temperature, pressure, etc.);
- simulate the fluid dynamics of the fuel-air mixture, from the time it is injected into the combustion chamber, through the burning of the fuel and consequent alteration of its chemical composition, temperature, and pressure, to the time it is exhausted from the chamber; and
- simulate the chemical processes involved in the combustion of the fuel, which for fuels such as *n*-heptane (a model for gasoline) can involve hundreds of chemical species and thousands of reactions.

The problem with the latter is that many of the molecular species involved in combustion have not yet been observed and many of the reactions have not yet been characterized (reaction energetics and rates) in the laboratory. From this it is clear that supercomputers will play a critical role in understanding the behavior of complex systems. It is simply not possible for scientists to understand how all of these physical and chemical processes interact to determine the behavior of an internal combustion engine without computers to handle all of the bookkeeping. It will not be possible for experimental chemists to identify and characterize all of the important chemical species and reactions of importance in flames; reliable computational predictions will be essential to realizing the goal.

An example of cutting-edge research on the fundamental chemical processes occurring in internal combustion engines is the work being carried out at the Combustion Research Facility of Sandia National Laboratories. For example, Jackie Chen and her group are working on the development of computational models to describe autoignition. Autoignition is the process that ignites a fuel by the application of heat, not a flame or spark. The fuel in diesel engines is ignited by autoignition; a spark plug is not present. Autoignition is also the basis of a very efficient internal combustion engine with extremely low emissions—the revolutionary homogeneous charge, compression ignition, or HCCI engine. The catch is that HCCI engines can not yet be used in cars and trucks because it is not yet possible to properly control the "flameless" combustion of fuels associated with autoignition.

Figure 12 illustrates recent results from J. Chen and T. Echekki at Sandia National Laboratories (*unpublished*). This figure plots the concentration of peroxyl radical (HOO), which has been found to be an indicator of the onset of autoignition, as a function of time in a H_2-air mixture from two-dimensional, direct numerical simulations. These results led Chen and Echekki to identify new chemical pathways for autoignition and led them to propose a new approach to

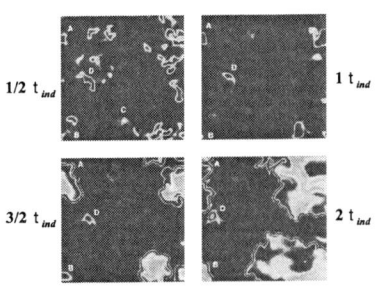

Evolution of hydroperoxy, HO_2, at different fractions of autoignition induction time.

- Turbulent mixing strongly affects ignition delay by changing chemical branching/termination balance
- New models may assist innovative engine design (HCCI)
- Limited by existing computer capabilities

FIGURE 12 Coupling chemistry and fluid dynamics: H_2 autoignition in turbulent mixtures. Courtesy of J. H. Chen and T. Echekki, Sandia National Laboratories.

describing autoignition in terms of relevant flow and thermochemical parameters. The simulations show that autoignition is initiated in discrete kernels or structures that evolve differently, depending strongly upon the exact local environment in which they occur. This is easily seen in the figure. But, this work is only a beginning. Turbulent fluctuations as well as the autoignition kernels are inherently three dimensional, and more complex fossil fuels will undoubtedly give rise to new behaviors that must be studied in a full range of environments to gain the scientific insight needed to develop predictive models of real-world autoignition systems. However, the computing requirements for such calculations require software and hardware capable of sustaining tens of teraflops.

An internal combustion engine is just one example of a complex system. There will be many opportunities to use our fundamental knowledge of chemistry to better understand complex, real-world systems. Other examples include a wide variety of processes involved in industrial chemical production, the molecular processes that determine cellular behavior, and the chemical processes affecting the formation of pollutants in urban environments. At the workshop, Jim Heath made a presentation on nanochemistry, *i.e.*, nanoscale molecular processes. One of the most intriguing aspects of nanoscale phenomena is that nanoscale systems are the first systems to exhibit multiple scales in chemistry—the molecular scale and the nanoscale. Phenomena at the molecular scale are characterized by time scales of femto- to picoseconds and distance scales of tenths of a nanometer to a nanometer. Nanoscale phenomena on the other hand often operate on micro- to millisecond (or longer) time scales and over distances of 10-100 (or more) nm. So, in addition to whatever practical applications nanochemistry has, it also represents a opportunity to understand how to describe disjoint temporal and spatial scales in computational modeling. Achieving this goal will require close collaboration between experimental chemists synthesizing and characterizing nanochemical systems and computational chemists modeling these systems. The concepts developed in such studies may be applicable to scaling to the meso- and macroscales.

Data Storage, Mining, and Analysis in Genomics and Related Sciences

Let me turn now to a very different type of problem—the data problem in science. The research universities in North Carolina, like most major research universities in the United States, want to become major players in the genomics revolution. This revolution not only portends a major advance in our understanding of the fundamentals of life, but also promises economic rewards to those states that make the right investments in this area. However, the problem with the "new biology" is that it is very different than the "old biology." It is not just more quantitative: it has rapidly become a very data-intensive activity. It is not data intensive now—most of the data currently available can be stored and analyzed on personal computers or small computer–data servers. But, with the quantity of data increasing exponentially, the amount of data will soon overwhelm all of those institutions that have not built the information infrastructure needed to manage it. Few universities are doing this. Most are simply assuming that their faculty will solve the problem on an individual basis. I don't believe this will work—the problem is of such a scale that it cannot be addressed by point solutions. Instead, to be a winner in the genomics sweepstakes, universities or even university consortia must invest in the integrated information infrastructure that will be needed to store, mine, and analyze the new biological data.

The next two figures illustrate the situation. Figure 13 is a plot of the number of gigabases (*i.e.*, linear sequences of A's, T's, C's, and G's) stored in GenBank from 1982 to the present.[25] As can be seen, the number of gigabases in GenBank was negligible until about 1991, although I can assure you that scientists were hard at work from 1982 to 1991 sequencing DNA; the technology simply did not permit large-scale DNA sequencing and thus the numbers don't show on the plot. The inadequacy of the technology used to sequence DNA was widely recognized when the Office of Science in the U.S. Department of Energy initiated its human DNA sequencing program in 1986 and substantial investments were made in the development of high-throughput sequencing technologies. When the Human Ge-

---

[25]There are three database respositories that store and distribute genome nucleotide sequence data. GenBank is the repository in the U.S. and is managed by the NIH (see: *http://www.ncbi.nlm.nih.gov*); DDBJ is the DNA Data Bank of Japan managed by the Japanese National Institute of Genetics (see: *http://www.nig.ac.jp/home.html*); and the EMBL Nucleotide Database is maintained by the European Bioinformatics Institute (see: *http://www.ebi.ac.uk/*). Sequence information is exchanged between these sites on a daily basis.

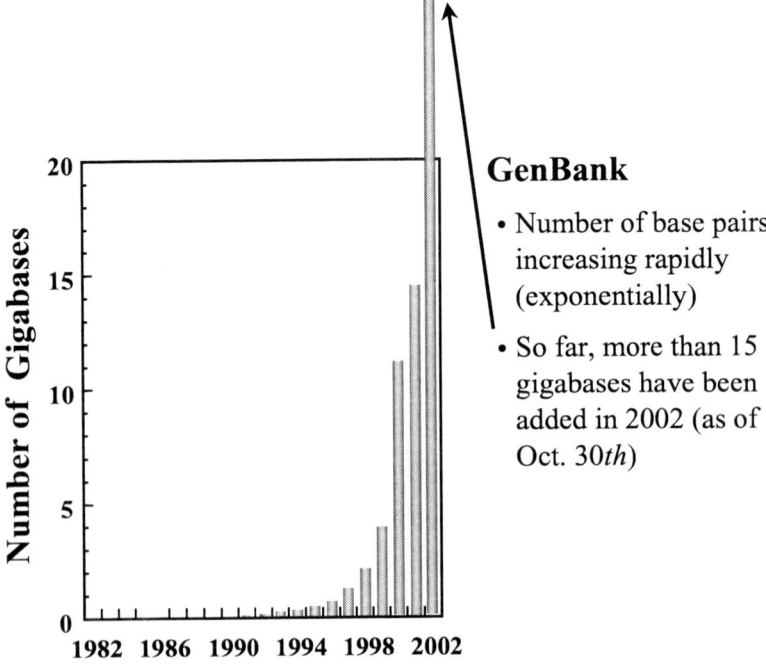

FIGURE 13 Background: exponential growth in GenBank.

nome Project, a joint effort of the National Institutes of Health and the Office of Science, began in 1990, a similar approach was followed. We are now reaping the benefits of those investments. At the present time, the doubling time for GenBank is less than one year and decreasing—in the first 10 months of 2002, more sequences were added to GenBank than in all previous years.

The other "problem" that must be dealt with in genomics research is the diversity of databases (Figure 14). There is not just one database of importance to molecular biologists involved in genomics, proteomics, and bioinformatics research; there are many, each storing a specific type of information. Each year the journal *Nucleic Acids Research* surveys the active databases in molecular biology and publishes information on these databases as the first article of the year. In January 2002, 335 databases were identified. Many of these databases are derived from GenBank; others are independent of GenBank. Whatever the source, however, all are growing rapidly. The challenge to the researcher in genomics is to link the data in these databases with the data that he/she is producing to advance the understanding of biological structure and function. Given the size and diversity of the databases, this is already intractable without the aid of a computer.

Putting all of the above together, we find ourselves in a most unusual situa-

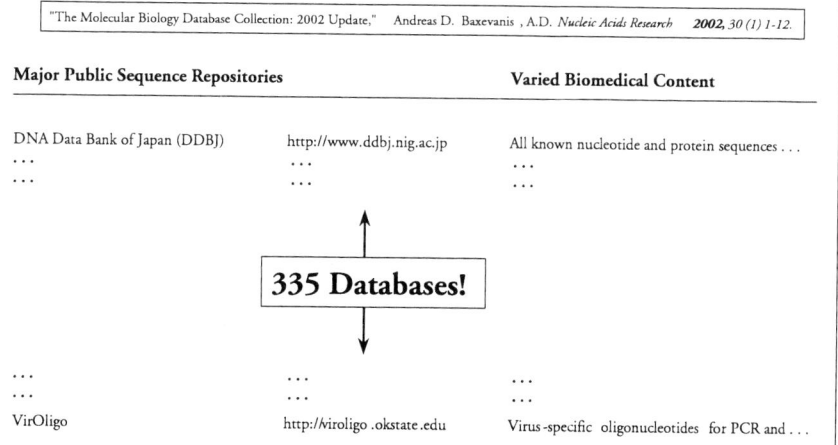

FIGURE 14 Background: number and diversity of databases.

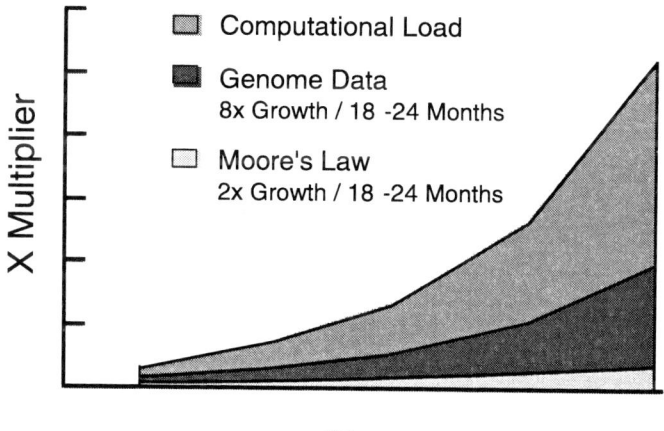

FIGURE 15 Background: mining and analysis of genomic data. Courtesy of TimeLogic Corporation.

tion in biology (Figure 15, from TimeLogic[26]). Moore's Law is represented at the bottom of the graph; this represents a doubling of computing capability every 18-24 months. The middle band illustrates the rate at which the quantity of genomic data is increasing. Note that, as seen above, it outstrips the growth in com-

---

[26]See: *http://www.timelogic.com/*.

puting power from Moore's Law. The upper band represents the rate of growth in computing resources needed for genomic research (i.e., the amount of computing power that will be needed to mine and analyze genomic data). Clearly, the day of reckoning is coming. Although most biologists involved in genomic and related research are very comfortable with the situation they are currently in, when storage and computing requirements are increasing at an exponential rate, one can be quite comfortable one year and absolutely miserable the next.

Of course, those biologists most heavily involved in genomics research understand that substantial computing power and data storage capabilities will be needed to support their research. Even so, many still underestimate the growing magnitude of the problem. Faculty in Duke University, North Carolina State University, and the University of North Carolina at Chapel Hill are attempting to solve the problem by building PC-based clusters (usually Linux clusters) to store and analyze their data. There are many such clusters all over the three universities with more to come. If there ever was an argument for collective action, this is it. Not only will it be difficult to grow the clusters at the pace needed to keep up with the proliferation of data, but it will be difficult to house and operate these machines as well as provide many essential services needed to ensure the integrity of the research (e.g., data backup and restore).

One difficulty with taking collective action is that each researcher wants to have his/her data close at hand, but these data need to be seamlessly integrated with all of the other data that are available, both public data being generated in the universities in North Carolina and data being stored in databases all over the world. In addition, the researchers want to be confident than their private data are secure. One way to achieve this goal is to use Grid technologies. With Grid technologies, a distributed set of computing and data storage resources can be combined into a single computing and data storage system. In fact, my idea was to spread this capability across the whole state of North Carolina. All of the 16 campuses of the University of North Carolina plus Duke University and Wake Forest University would be tied together in a Grid and would have ready access to the computing and data storage resources available throughout the Grid whether they were physically located in the Research Triangle, on the coast, or in the mountains. Although the biggest users of such a Grid would likely be the research universities, all of the universities need to have access to this capability if they are to properly educate the next generation of biologists.

Some of the attributes of the North Carolina Bioinformatics Grid[27] are summarized in Figure 16. At the top of the list are general capabilities such as single sign-on (a user need only log onto the Grid; thereafter all data and computing resources to which they are entitled become available) and a system-wide approach to security rather than a university-by-university approach. The latter is particularly important in a multicampus university. Other capabilities include fine

---

[27] For more information on the NC BioGrid, see: *http://www.ncbiogrid.org*.

Attributes

- Single sign-on, system-wide security
- Policy-based resource sharing

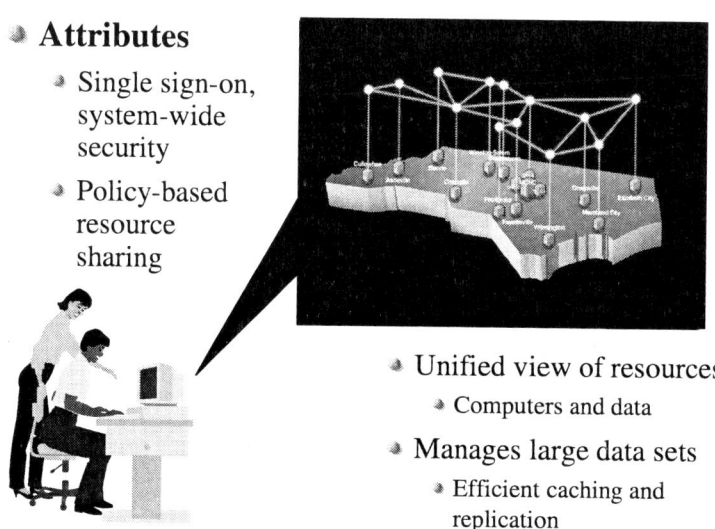

- Unified view of resources
  - Computers and data
- Manages large data sets
  - Efficient caching and replication

FIGURE 16 The North Carolina Bioinformatics Grid.

grain control of resource sharing based on policies set at the university and campus level (controlling who has access to which resources for what length of time). This not only protects the rights of the individual campuses, but allows the performance of the BioGrid to be optimized by siting computing and data storage resources at the optimal locations on the network. The Grid software underlying the BioGrid provides a unified view of all of the resources available to a user whether they are located on his/her campus or distributed across the state. In fact, one of the chief goals of Grids, such as the BioGrid, is to simplify the interaction of scientists with computers, data stores, and networks, allowing them to focus on their scientific research rather than the arcana of computing, data management, and networking. Finally, Grids, unlike other attempts at distributed computing such as NFS and AFS, are designed to efficiently handle large datasets. At times, to optimize performance, the Grid management software may decide to cache a dataset locally (e.g., if a particular dataset is being heavily used by one site) or replicate a dataset, if it is being heavily used on an ongoing basis.

There are many economies and benefits that will be realized with the NC BioGrid. For example, at the present time all of the universities maintain their own copies of some of the public databases. Although this is not a significant expense now, it certainly will be when these databases hold petabytes of data rather than hundreds of gigabytes of data. In addition to the hardware required to store and access such massive amounts of data, a seasoned staff will be required to manage such a resource. Important data management services such as backup and restore of research data can be assured more easily in a Grid environment

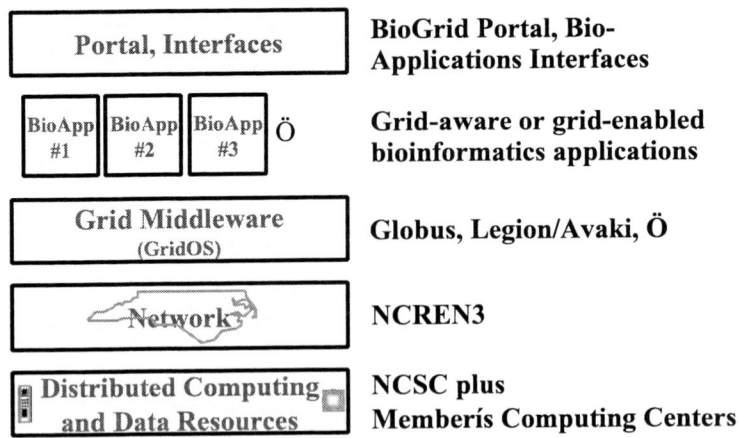

FIGURE 17 Elements of the North Carolina BioGrid.

than in a loosely coupled system of computers. Finally, economies will be realized by allowing each campus to size its computing and data storage system to meet its average workload and using resource sharing arrangements with the other campuses on the BioGrid to offload its higher-than-average demands.

So, how is a BioGrid constructed? Like many applications in computing, a Grid is best viewed as a layered system (Figure 17). At the lowest level are the distributed computing and data resources. It is assumed that these resources are geographically distributed, although one particular site may be dominant. This is the case in North Carolina where the dominant site will be the North Carolina Supercomputing Center (terascale supercomputer and petascale data store); yet significant resources will also be needed at the three major campuses, and some level of resources will be needed at all campuses. The next level is the network that links the distributed computing and data storage resources. If the datasets are large, a high-performance network is needed. North Carolina is fortunate is having a statewide, high-performance network—the North Carolina Research and Education Network (NCREN). This network serves all 16 campuses of the University of North Carolina plus Duke University and Wake Forest University and is currently being upgraded to ensure that no university is connected to NCREN at less than OC-3 (155 Mbps) and many will be connected at OC-12 (622 Mbps). The next layer is the operating system for the Grid, something called Grid middleware. It is the middleware that adds the intelligence associated with the Grid to the network. There are currently two choices for the Grid operating system, Globus[28] and Legion-Avaki.[29] Although Globus, which was first developed by researchers at

---

[28] http://www.globus.org.

Argonne National Laboratory and the University of Southern California, currently has the greatest mind-share, it is a toolkit for building Grids and requires a considerable amount of expertise to set up a Grid. Avaki, a commercial product that grew out of Legion, a Grid environment developed at the University of Virginia, provides an all-in-one system. The North Carolina BioGrid Testbed has both Grid operating systems running. Both Globus and Avaki are currently migrating to open standards (Open Grid Services Infrastructure and Open Grid Services Architecture[30]).

The top two layers of the BioGrid are what the user will see. They consist of applications such as BLAST, CHARMM, or Gaussian, modified to make them aware of the services that they can obtain from the Grid, and the portals and web interfaces that allow access to these applications and data resources in a user-friendly way. Portals can also provide access to workflow management services. Often a scientist doesn't just perform an isolated calculation. In genomics research, a typical "computer experiment" might involve comparing a recently discovered sequence with the sequences available in a number of databases (BLAST calculation). If related sequences are found in these databases, then the scientist may wish to know if three-dimensional molecular structures exist for any of the proteins coded by these related sequences (data mining) or may want to know what information is available on the function of these proteins in their associated biological systems (data mining). And, the list goes on. The development of software that uses computers to manage this workflow, eliminating the time that the scientist has to spend massaging the output from one application to make it suitable for input to the next application, is a major opportunity for advancing research in molecular biology.

When we announced that we were going to build a BioGrid in North Carolina, we were immediately approached by several biology-related software companies stating that we didn't have to build a BioGrid, they already had one. After parrying several such thrusts, Chuck Kesler at MCNC decided that we needed to more carefully define what we meant by a Grid. This is done in Figure 18. Although the companies were able to provide bits and pieces of the functionality represented in this figure, none were able to provide a significant fraction of these capabilities.

So, this is a Grid. I challenge the participants in this workshop to think about applications for Grid technologies in chemical science and technology. Clearly, chemists and chemical engineers are not yet confronted with the flood of data that is becoming available in the genomics area but the amount of data that we have in chemistry is still substantial. Ann Chaka discusses the issue of chemical data storage and management in her paper.

---

[29]*http://www.avaki.com.*
[30]*http://www.gridforum.org/Documents/Drafts/default.htm.*

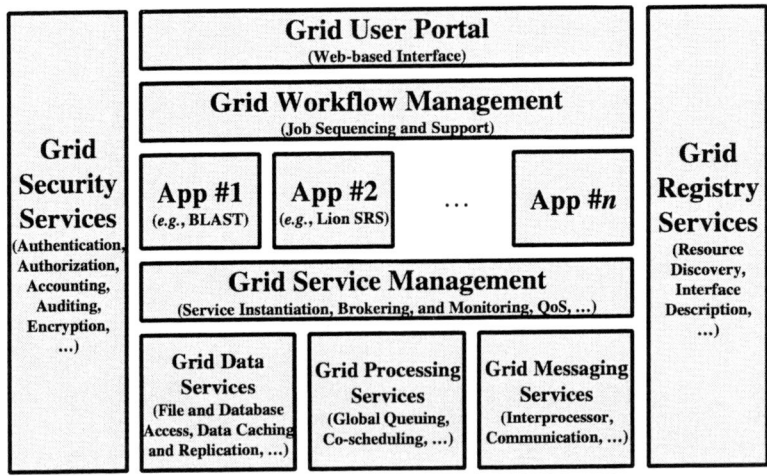

FIGURE 18 No, what you have is not a Grid!

Virtual Laboratories in Chemical Science and Technology

The last topic that I want to discuss is the concept of virtual laboratories. Advances in networking and computing have made remote access to instruments not only possible but (reasonably) convenient, largely removing the inconveniences associated with distance (e.g., travel to the remote instrument site, specified time slots for instrument availability, limited time to consult with site scientists on experimental details, etc.). The concept of virtual laboratories is quite advanced in astronomy, where the instruments (large telescopes) are very expensive and often located in remote regions (on the tops of mountains scattered all over the world). In chemical science and technology, on the other hand, remote access to instruments is largely a foreign concept. However, as instruments for chemical research continue to increase in cost (e.g., the highest field NMRs and mass spectrometers currently available already cost several million dollars), it will soon become desirable, if not necessary, for universities and research laboratories to share these instruments rather than buy their own. Investments in software and hardware to enable remote access to these instruments can dramatically decrease the barriers associated with the use of instruments located at a remote site.

One of the major investments that I made as associate director for computing and information science (and later as Director) in the Environmental Molecular Sciences Laboratory at Pacific Northwest National Laboratory was in the development of collaboratories, which were defined by Wm. A. Wulf as[31]

---

[31]Wulf, Wm. A. *The National Collaboratory: A White Paper in Towards a National Collaboratory;* unpublished report of a NSF workshop, Rockefeller University, NY. March 17-18, 1989.

a "center without walls," in which the nation's researchers can perform their research without regard to geographical location—interacting with colleagues, accessing instrumentation, sharing data and computational resources, and accessing information in digital libraries.

(See also *National Collaboratories: Applying Information Technologies for Scientific Research*[32]). I did this because PNNL is located in the Pacific Northwest, a substantial distance from most of the major centers of research in the United States. Since EMSL is a National User Facility, much like the Advanced Photon Source at Argonne National Laboratory or the Advanced Light Source at Lawrence Berkeley National Laboratory, we needed to minimize the inconvenience for scientists wishing to use EMSL's facilities to support their research programs in environmental molecular science. Allowing scientists access to EMSL's resources via the Internet was key to realizing this goal.

In EMSL there are a number of first-of-a-kind and one-of-a-kind instruments that could be made available to the community via the Internet. We decided to focus on the instruments in the High-Field Nuclear Magnetic Resonance Facility as a test case. When we began, this facility included one of the first 750-MHz NMRs in the United States; it now includes an 800-MHz NMR, and a 900-MHz NMR is expected to come on-line shortly. The NMRs were very attractive candidates for this test since the software for field servicing the NMRs provided much of the basic capability needed to make them accessible over the Internet.

Development of the Virtual NMR Facility (VNMRF, Figure 19) was a true collaboration between NMR spectroscopists and computer scientists. One of the first discoveries was that simply implementing secure remote control of the NMRs was not sufficient to support the VNMRF. Many other capabilities were needed to realize the promise of the VNMRF, including

- real-time videoconferencing,
- remotely controlled laboratory cameras, and
- real-time computer displays sharing a Web-based electronic laboratory notebook and other capabilities.

Of particular importance was the ability of remote researchers to discuss issues related to the experiment with collaborators, scientists, and technicians in the EMSL before, during and after the experiment as well as to work together to analyze the results of the experiment. These discussions depended on the availability of electronic tools such as the Electronic Laboratory Notebook and the Televiewer (both are included in PNNL's CORE2000 software package[33] for supporting collaborative research).

---

[32]*National Collaboratories: Applying Information Technologies for Scientific Research*, National Research Council, National Academy Press, Washington, DC, 1993.
[33]*http://www.emsl.pnl.gov:2080/docs/collab/*.

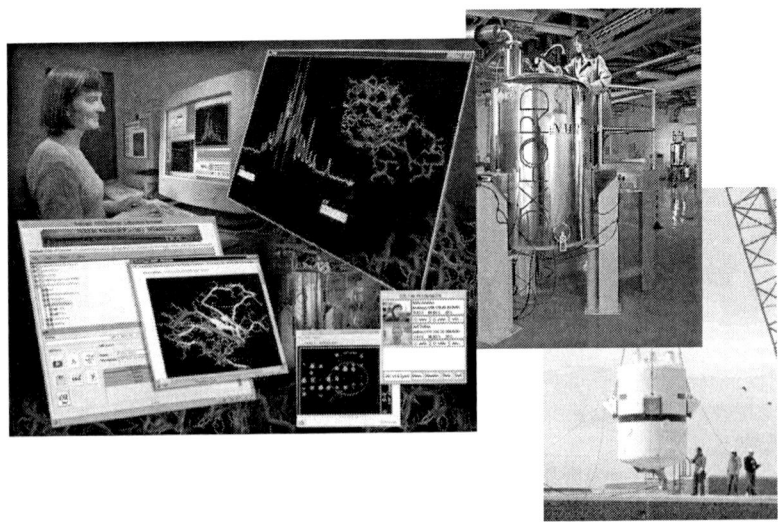

FIGURE 19 The Virtual NMR Facility at the Environmental Molecular Sciences Laboratory (Pacific Northwest National Laboratory).

Using the software developed or integrated into CORE2000 by EMSL staff, a remote scientists can schedule a session on an NMR, discuss the details of the experiment with EMSL staff, send the sample to EMSL, watch the technician insert the sample into the NMR, control the experiment using a virtual control panel, and visualize the data being generated in the experiment. The scientists can do everything from the virtual control panel displayed on his/her computer display that he/she can do sitting at the real control panel. Currently, over half of the scientists using EMSL's high-field NMRs use them remotely, a testimony to the effectiveness of this approach to research in chemical science and technology. A more detailed account of the VNMRF can be found in the paper by Keating et al.[34] See also the VNMRF home page.[35]

There are a few other virtual facilities now in operation. A facility for electron tomography is operational at the University of California at San Diego[36] and has proven to be very successful. There are electron microscopy virtual laborato-

---

[34]K. A. Keating, J. D. Myers, J. G. Pelton, R. A. Bair, D. E. Wemmer, and P. D. Ellis, *J. Mag. Res.* **2000**, *143*, 172-183.
[35]*http://www.emsl.pnl.gov:2080/docs/collab/virtual/EMSLVNMRF.html.*
[36]*http://ncmir.ucsd.edu/Telescience.*

ries in operation at Argonne National Laboratory[37] and Oak Ridge National Laboratory.[38] Despite these successes, however, full-fledged support for virtual laboratories has not yet materialized. Those federal agencies in charge of building state-of-the-art user facilities for science and engineering rarely consider the computing, data storage, and networking infrastructure that will be needed to make the facility widely accessible to the scientific community. Even less do they consider the investments in software development and integration that will be needed to provide this capability. This is a lost opportunity that results in much wasted time on the part of the scientists who use the facility to further their research programs.

Conclusion

I hope that this paper has given you an appreciation of the potential impact of advances in computing, data storage, and networking in chemical science and technology. These advances will profoundly change our field. They will greatly enhance our ability to model molecular structure, energetics, and dynamics, providing insights into molecular behavior that would be difficult, if not impossible, to obtain from experimental studies alone. They will also allow us to begin to model many complex systems in which chemical processes are an integral part (e.g., gasoline and diesel engines, industrial chemical production processes, and even the functioning of a living cell). The insights gained from these studies not only will deepen our understanding of the behavior of complex systems, but also will have enormous economic benefits.

The advances being made in Grid technologies and virtual laboratories will enhance our ability to access and use computers, chemical data, and first-of-a-kind or one-of-a-kind instruments to advance chemical science and technology. Grid technologies will substantially reduce the barrier to using computational models to investigate chemical phenomena and to integrating data from various sources into the models or investigations. Virtual laboratories have already proven to be an effective means of dealing with the rising costs of forefront instruments for chemical research by providing capabilities needed by researchers not co-located with the instruments—all we need is a sponsor willing to push this technology forward on behalf of the user community.

The twenty-first century will indeed be an exciting time for chemical science and technology.

---

[37]http://www.amc.anl.gov/.
[38]*http://www.ms.ornl.gov/htmlhome/mauc/MAGrem.htm.*

SYSTEMS APPROACHES IN BIOINFORMATICS AND COMPUTATIONAL GENOMICS

Christodoulos A. Floudas
Princeton University

The genomics revolution has generated a plethora of challenges and opportunities for systems approaches in bioinformatics and computational genomics. The essential completion of several genome projects, including that of the human genome, provided a detailed map from the gene sequences to the protein sequences. The gene sequences can be used to assist and/or infer the connectivity within or among the pathways. The overwhelmingly large number of generated protein sequences makes protein structure prediction from the amino acid sequence of paramount importance. The elucidation of the protein structures through novel computational frameworks that complement the existing experimental techniques provides key elements for the structure-based prediction of protein function. These include the identification of the type of fold, the type of packing, the residues that are exposed to solvent, the residues that are buried in the core, the highly conserved residues, the candidate residues for mutations, and the shape and electrostatic properties of the fold. Such elements provide the basis for the development of approaches for the location of active sites; the determination of structural and functional motifs; the study of protein-protein and protein-ligand complexes and protein-DNA interactions; the design of new inhibitors; and drug discovery through target selection, lead discovery and optimization. Better understanding of the protein-ligand and protein-DNA interactions will provide important information for addressing key topology related questions in both the cellular metabolic and signal transduction networks. In this paper, we discuss two components of the genomics revolution roadmap: (1) sequence to structure, and (2) structure to function. In the first, after a brief overview of the contributions, we present ASTRO-FOLD, which is a novel ab initio, approach for protein structure prediction. In the second, we discuss the approaches for de novo protein design and present an integrated structural, computational, and experimental approach for the de novo design of inhibitors for the third component of complement, C3. We conclude with a summary of the advances and a number of challenges.

Sequence to Structure: Structure Prediction in Protein Folding

Structure prediction of polypeptides and proteins from their amino acid sequences is regarded as a *holy grail* in the computational chemistry and molecular and structural biology communities. According to the *thermodynamic hypothesis*[1] the native structure of a protein in a given environment corresponds to the

---

[1] Anfinsen, C. B. *Science* **1973**, *181*, 223.

global minimum free energy of the system. In spite of pioneering contributions and decades of effort, the ab initio prediction of the folded structure of a protein remains a very challenging problem. The existing approaches for the protein structure prediction can be classified as : (1) homology or comparative modeling methods,[2,3,4,5,6] (2) fold recognition or threading methods,[7,8,9,10,11,12,13] (3) ab initio methods that utilize knowledge-based information from structural databases (e.g., secondary and/or tertiary structure restraints).[14,15,16,17,18,19,20] and (4) ab initio methods without the aid of knowledge-based information.[21,22,23,24,25,26,27]

In the sequel, we introduce the novel ASTRO-FOLD approach for the ab initio prediction of the three-dimensional structures of proteins. The four stages of the approach are outlined in Figure 1. The first stage involves the identification of helical segments[24] and is accomplished by partitioning the amino acid sequence into overlapping oligopeptides (e.g., pentapeptides, heptapeptides, nonapeptides).

---

[2]Bates, P. A.; Kelley, L. A.; MacCallum, R. M.; Sternberg, M. J. E. *Proteins* **2001**, *S5*, 39-46.
[3]Shi, J. Y.; Blundell, T. L.; Mizuguchi, K. *J. Mol. Biol.* **2001**, *310*, 243-257.
[4]Sali, A.; Sanchez, R. *Proc. Natl. Acad. Sci. U.S.A.* **1998**, *95*, 13597-13602.
[5]Fischer, D. *Proteins* **1999**, *S3*.
[6]Alazikani, B., Sheinerman, F. B.; Honig, B. *Proc. Natl. Acad. Sci. U.S.A.* **2001**, *98*, 14796-14801.
[7]Fischer, D.; Eisenberg, D. *Proc. Natl. Acad. Sci. U.S.A.* **1997**, *94*, 11929-11934.
[8]Skolnick, J.; Kolinski, A. *Adv. Chem. Phys.* **2002**, *120*, 131-192.
[9]McGuffin, L. J.; Jones, D. T. *Proteins* **2002**, *48*, 44-52.
[10]Panchenko, A. R.; Marchler-Bauer, A.; Bryant, S. H. *J. Mol. Biol.* **2000**, *296*, 1319-1331.
[11]Jones, D. T. *Proteins* **2001**, *S5*, 127-132.
[12]Skolnick, J.; Kolinski, A.; Kihara, D.; Betancourt, M.; Rotkiewicz, P.; Boniecki, M. *Proteins* **2001**, *S5*, 149-156.
[13]Smith, T. F.; LoConte, L.; Bienkowska, J.; Gaitatzes, C.; Rogers, R. G.; Lathrop, R. *J. Comp. Biol.* **1997**, *4*, 217-225.
[14]Ishikawa, K.; Yue, K.; Dill, K. A. *Prot. Sci.* **1999**, *8*, 716-721.
[15]Pedersen, J. T.; Moult, J. *Proteins* **1997**, *S1*, 179-184.
[16]Eyrich, V. A.; Standley, D. M.; Friesner, R. A. *Adv. Chem. Phys.* **2002**, *120*, 223-264.
[17]Xia, Y.; Huang, E. S.; Levitt, M.; Samudrala, R. *J. Mol. Biol.* **2000**, *300*, 171-185.
[18]Standley, D. M.; Eyrich, V. A.; An, Y.; Pincus, D. L.; Gunn, J. R.; Friesner, R. A. *Proteins* **2001**, *S5*, 133-139.
[19]Standley, D. M.; Eyrich, V. A.; Felts, A. K.; Friesner, R. A.; McDermott, A. E. *J. Mol. Biol.* **1999**, *285*, 1691-1710.
[20]Eyrich, V.; Standley, D. M.; Felts, A. K.; Friesner, R. A. *Proteins* **1999**, *35*, 41.
[21]Pillardy, J.; Czaplewski, C.; Liwo, A.; Wedemeyer, W. J.; Lee, J.; Ripoll, D. R.; Arlukowicz, P.; Oldziej, S.; Arnautova, Y. A.; Scheraga, H. A. *J. Phys. Chem. B* **2001**, *105*, 7299-7311.
[22]Pillardy, J.; Czaplewski, C.; Liwo, A.; Lee, J.; Ripoll, D. R.; Kazmierkiewicz, R.; Oldziej, S.; Wedemeyer, W. J.; Gibson, K. D.; Arnautova, Y. A.; Saunders, J.; Ye, Y. J.; Scheraga, H. A. *Proc. Natl. Acad. Sci. U.S.A.* **2001**, *98*, 2329-2333.
[23]Srinivasan, R.; Rose, G. D. *Proteins* **2002**, *47*, 489-495.
[24]Klepeis, J. L.; Floudas, C. A. *J. Comp. Chem.* **2002**, *23*, 245-266.
[25]Klepeis, J. L.; Floudas, C. A. *J.Comp. Chem.* **2003**, *24*, 191-208.
[26]Klepeis, J. L.; Floudas, C. A. *J. Global Optim.* unpublished.
[27]Klepeis, J. L.; Schafroth, H. D.; Westerberg, K. M.; Floudas, C. A. *Adv. Chem. Phys.* **2002**, *120*, 254-457.

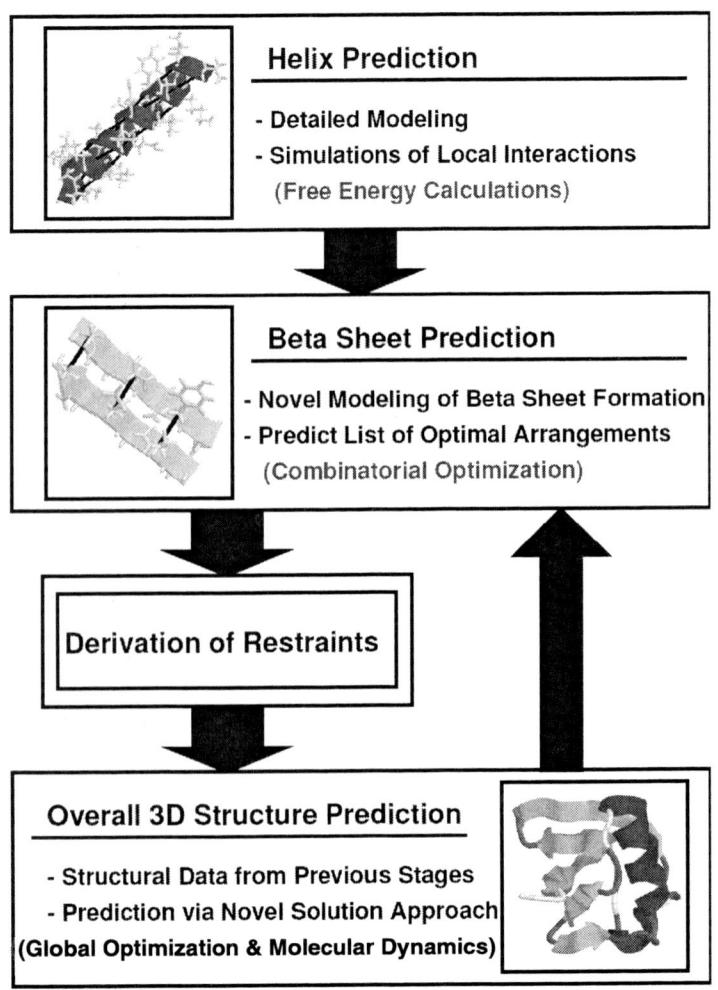

FIGURE 1 Overall flow chart for the ab initio structure prediction using ASTRO-FOLD.

The concept of partitioning the amino acid sequence into overlapping oligopeptides is based on the idea that helix nucleation relies on local interactions and positioning within the overall sequence. This is consistent with the observation that local interactions extending beyond the boundaries of the helical segment retain information regarding conformational preferences.[28] The partitioning pattern is generalizable and can be extended to heptapeptides, nonapeptides,

---

[28]Baldwin, R. L.; Rose, G. D. *TIBS* **1999**, *24*, 77-83.

or larger oligopeptides.[29] The overall methodology for the ab initio prediction of helical segments encompasses the following steps:[24] The overlapping oligopeptides are modeled as neutral peptides surrounded by a vacuum environment using the ECEPP/3 force field.[30] An ensemble of low potential energy pentapeptide conformations, along with the global minimum potential energy conformation, is identified using a modification of the $\alpha\beta\beta$ global optimization approach[31] and the conformational space annealing approach.[32] For the set of unique conformers Z, free energies (F^{har}_{vac}) are calculated using the harmonic approximation for vibrational entropy.[31] The energy for cavity formation in an aqueous environment is modeled using a solvent-accessible surface area expression $F_{cavity} = \gamma A + b$, where A is the surface area of the protein exposed to the solvent. For the set of unique conformers Z, the total free energy F_{total} is calculated as the summation of F^{har}_{vac}, F_{cavity}, and F_{solv}, which represents the difference in polarization energies caused by the transition from a vacuum to a solvated environment, and F_{ionize}, which represents the ionization energy. The calculation of F_{solv} and F_{ionize} requires the use of a Poisson-Boltzmann equation solver.[33] For each oligopeptide, total free energy values (F_{total}) are used to evaluate the equilibrium occupational probability for conformers having three central residues within the helical region of the ϕ-Ψ space. Helix propensities for each residue are determined from the average probability of those systems in which the residue in question constitutes a core position.

In the second stage, β strands, β sheets, and disulfide bridges are identified through a novel superstructure-based mathematical framework originally established for chemical process synthesis problems.[25,34] Two types of superstructure are introduced, both of which emanate from the principle that hydrophobic interactions drive the formation of structure. The first one, denoted as *hydrophobic residue-based superstructure*, encompasses all potential contacts between pairs of hydrophobic residues (i.e., a contact between two hydrophobic residues may or may not exist) that are not contained in helices (except cystines, which are allowed to have cystine-cystine contacts even though they may be in helices). The second one, denoted as β-*strand-based superstructure*, includes all possible β-strand arrangements of interest (i.e., a β strand may or may not exist) in addition to the potential contacts between hydrophobic residues. The hydrophobic residue-based and β-strand-based superstructures are formulated as mathematical

---

[29]Anfinsen, C.; Scheraga, H. *Adv. Prot. Chem.* **1975**, *29*, 205.

[30]Némethy, G.; Gibson, K. D.; Palmer, K. A.; Yoon, C. N.; Paterlini, G.; Zagari, A.; Rumsey, S.; Scheraga, H. A. *J. Phys. Chem.* **1992**, *96*, 6472.

[31]Klepeis, J. L.; Floudas, C. A. *J. Chem. Phys.* **1999**, *110*, 7491-7512.

[32]Lee, J.; Scheraga, H.; Rackovsky, S. *Biopolymers* **1998**, *46*, 103.

[33]Honig, B.; Nicholls, A. *Science* **1995**, *1*, 11144-1149.

[34]Floudas, C. A. *Nonlinear and Mixed-Integer Optimization;* Oxford University Press: New York, NY, 1995.

models that feature three types of binary variables: (1) representing the existence or nonexistence of contacts between pairs of hydrophobic residues; (2) denoting the existence or nonexistence of the postulated β strands; and (3) representing the potential connectivity of the postulated β strands. Several sets of constraints in the model enforce physically legitimate configurations for antiparallel or parallel β strands and disulfide bridges, while the objective function maximizes the total hydrophobic contact energy. The resulting mathematical models are integer linear programming (ILP) problems that not only can be solved to global optimality, but also can provide a rank-ordered list of alternate β-sheet configurations.[25]

The third stage determines pertinent information from the results of the previous two stages. This involves the introduction of lower and upper bounds on dihedral angles of residues belonging to predicted helices or β strands, as well as restraints between the Cα atoms for residues of the selected β-sheet and disulfide-bridge configuration. Furthermore, for segments that are not classified as helices or β strands, free-energy runs of overlapping heptapeptides are conducted to identify tighter bounds on their dihedral angles.[24,27,35]

The fourth stage of the approach involves the prediction of the tertiary structure of the full protein sequence.[26] Formulation of the problem relies on the minimization of the energy using a full atomistic force field, ECEPP/3[30] and on dihedral angle and atomic distance restraints acquired from the previous stage. To overcome the multiple minima difficulty, the search is conducted using the αββ global optimization approach, which offers theoretical guarantee of convergence to an ε-global minimum for nonlinear optimization problems with twice-differentiable functions.[27,36,37,38,39]

This global optimization approach effectively brackets the global minimum by developing converging sequences of lower and upper bounds, which are refined by iteratively partitioning the initial domain. Upper bounds correspond to local minima of the original nonconvex problem, while lower bounds belong to the set of solutions of convex lower bounding problems, which are constructed by augmenting the objective and constraint functions by separable quadratic terms. To ensure nondecreasing lower bounds, the prospective region to be bisected is required to contain the infimum of the minima of lower bounds. A nonincreasing sequence for the upper bound is maintained by selecting the minimum over all the previously recorded upper bounds. The generation of low-energy starting points

---

[35]Klepeis, J. L.; Pieja, M. T.; Floudas, C. A. *Comp. Phys. Comm.* **2003**, *151*, 121-140.
[36]Adjiman, C. S.; Androulakis, I. P.; Floudas, C. A. *Computers Chem. Engng.* **1998**, *22*, 1137-1158.
[37]Adjiman, C. S.; Androulakis, I. P.; Floudas, C. A. *Computers Chem. Engng.* **1998**, i 1159-1179.
[38]Adjiman, C. S.; Androulakis, I. P.; Floudas, C. A. *AIChE Journal* **2000**, *46*, 1769-1797.
[39]Floudas, C. A. *Deterministic Global Optimization: Theory, Methods and Applications, Nonconvex Optimization and its Applications*; Kluwer Academic Publishers: Dordecht, 2000.

for constrained minimization is enhanced by introducing torsion angle dynamics[40] within the context of the $\alpha\beta\beta$ global optimization framework.[26]
Two viewpoints provide competing explanations of the protein-folding question. The classical opinion regards folding as hierarchic, implying that the process is initiated by rapid formation of secondary structural elements, followed by the slower arrangement of the tertiary fold. The opposing perspective is based on the idea of a hydrophobic collapse and suggests that tertiary and secondary features form concurrently. ASTRO-FOLD bridges the gap between the two viewpoints by introducing a novel ab initio approach for tertiary structure prediction in which helix nucleation is controlled by local interactions, while non-local hydrophobic forces drive the formation of β structure. The agreement between the experimental and predicted structures (RMSD, root mean squared deviation: 4-6 Å for segments up to 100 amino acids) through extensive computational studies on proteins up to 150 amino acids reflects the promise of the ASTRO-FOLD method for generic tertiary structure prediction of polypeptides.

Structure to Function: De Novo Protein Design

The de novo protein design relies on understanding the relationship between the amino acid sequence of a protein and its three-dimensional structure.[41,42,43,44,45,46] *This problem begins with a known protein three-dimensional structure and requires the determination of an amino acid sequence compatible with this structure. At the outset the problem was termed the "inverse folding problem"*[46,47] *since protein design has intimate links to the well-known protein folding problem.*[48]

Experimentalists have applied the techniques of mutagenesis, rational design, and directed evolution[49,50] to the problem of protein design, and although these approaches have provided successes, the searchable sequence space is highly restricted.[51,52] Computational protein design allows for the screening of

---

[40]Güntert, P.; Mumenthaler, C.; Wüthrich, K. *J. Mol. Biol.* **1997**, *273*, 283-298.
[41]Ventura, S.; Vega, M.; Lacroix, E.; Angrand, I.; Spagnolo, L.; Serrano, L. *Nature Struct. Biol.* **2002**, *9*, 485-493.
[42]Neidigh, J. W.; Fesinmeyer, R. M.; Andersen, N. H. *Nature Struct. Biol.* **2002**, *9*, 425-430.
[43]Ottesen, J. J.; Imperiali, B. *Nature Struct. Biol.* **2001**, *8*, 535-539.
[44]Hill, R. B.; DeGrado, W. F. *J. Am. Chem. Soc.* **1998**, *120*, 1138-1145.
[45]Dahiyat, B. I.; Mayo, S. L. *Science* **1997**, *278*, 82-87.
[46]Drexler, K. E. *Proc. Natl. Acad. Sci. U.S.A.* **1981**, *78*, 5275-5278.
[47]Pabo, C. *Nature* **1983**, *301*, 200.
[48]C. Hardin, T. V. P.; Luthey-Schulten, Z. *Curr. Opin. Struc. Biol.* **2002**, *12*, 176-181.
[49]Bowie, J. U.; Reidhaar-Olson, J. F.; Lim, W. A.; Sauer, R. T. *Science* **1990**, *247*, 1306-1310.
[50]Moore, J. C.; Arnold, F. H. *Nat. Biotechnol.* **1996**, *14*, 458-467.
[51]DeGrado, W. F.; Wasserman, Z. R.; Lear, J. D. *Science* **1989**, *243*, 622-628.
[52]Hecht, M. H.; Richardson, D. S.; Richardson, D. C.; Ogden, R. C. *Science* **1990**, *249*, 884-891.

overwhelmingly large sectors of sequence space, with this sequence diversity subsequently leading to the possibility of a much broader range of properties and degrees of functionality among the selected sequences. Allowing for all 20 possible amino acids at each position of a small 50-residue protein results in 20^{50} combinations, or more than 10^{65} possible sequences. From this astronomical number of sequences, the computational sequence selection process aims at selecting those sequences that will be compatible with a given structure using efficient optimization of energy functions that model the molecular interactions. The first attempts at computational protein design focused only on a subset of core residues and explored steric van der Waals-based energy functions, although over time they evolved to incorporate more detailed models and interaction potentials. Once an energy function has been defined, sequence selection is accomplished through an optimization based search designed to minimize the energy objective. Both stochastic[53,54] and deterministic[55,56] methods have been applied to the computational protein design problem. Recent advances in the treatment of the protein design problem have led to the ability to select novel sequences given the structure of a protein backbone. The first computational design of a full sequence to be experimentally characterized was the achievement of a stable zinc-finger fold ($\beta\beta\alpha$) using a combination of a backbone-dependent rotamer library with atomistic-level modeling and a dead-end elimination-based algorithm.[45] Despite these breakthroughs, issues related to the stability and functionality of these designed proteins remain sources of frustration.

We have recently introduced a combined structural, computational, and experimental approach for the de novo design of novel inhibitors such as variants of the synthetic cyclic peptide Compstatin.[57] A novel two-stage computational protein design method is used not only to select and rank sequences for a particular fold but also to validate the stability of the fold for these selected sequences. To correctly select a sequence compatible with a given backbone template that is flexible and represented by several NMR structures, an appropriate energy function must first be identified. The proposed sequence selection procedure is based on optimizing a pairwise distance-dependent interaction potential. A number of different parameterizations for pairwise residue interaction potentials exist; the one employed here is based on the discretization of alpha carbon distances into a set of 13 bins to create a finite number of interactions, the parameters of which were derived from a linear optimization formulated to favor native folds over

---

[53]Wernisch, L.; Hery, S.; Wodak, S. J. *J. Mol. Biol.* **2000**, *301*, 713-736.
[54]Desjarlais, J. R.; Handel, T. M. *J. Mol. Biol.* **1999**, *290*, 305-318.
[55]Desmet, J.; Maeyer, M. D.; Hazes, B.; Lasters, I. *Nature* **1992**, *356*, 539-542.
[56]Koehl, P.; Levitt, M. *Nature Struct. Biol.* **1999**, *6*, 108.
[57]Klepeis, J. L.; Floudas, C. A.; Morikis, D.; Tsokos, C. G.; Argyropoulos, E.; Spruce, L.; Lambris, J. D. **2002**, submitted.

optimized decoy structures.[58,59] The resulting potential, which involves 2730 parameters, was shown to provide higher Z scores than other potentials and place native folds lower in energy.[58,59]

The formulation allows a set of mutations for each position i that in the general case comprises all 20 amino acids. Binary variables y_i^j and y_k^l can be introduced to indicate the possible mutations at a given position. That is, the y_i^j variable will indicate which type of amino acid is active at a position in the sequence by taking the value of 1 for that specification. The objective is to minimize the energy according to the amino acid pair and distance dependent energy parameters that multiply the binary variables. The composition constraints require that there is at most one type of amino acid at each position. For the general case, the binary variables appear as bilinear combinations in the objective function. This objective can be reformulated as a strictly linear (integer linear programming) problem.[57] The solution of the ILP problem can be accomplished rigorously using branch and bound techniques,[34] making convergence to the global minimum energy sequence consistent and reliable. Finally, for such an ILP problem it is straightforward to identify a rank-ordered list of the low-lying energy sequences through the introduction of integer cuts[34] and repetitive solution of the ILP problem.

Once a set of low-lying energy sequences has been identified via the sequence selection procedure, the fold validation stage is used to identify an optimal subset of these sequences according to a rigorous quantification of conformational probabilities. The foundation of the approach is grounded in the development of conformational ensembles for the selected sequences under two sets of conditions. In the first circumstance the structure is constrained to vary, with some imposed fluctuations, around the template structure. In the second case a free-folding calculation is performed for which only a limited number of restraints are likely to be incorporated, with the underlying template structure not being enforced. The distance constraints introduced for the template-constrained simulation can be based on the structural boundaries defined by the NMR ensemble, or simply by allowing some deviation from a subset of distances provided by the structural template; hence they allow for a flexible template on the backbone.

The formulations for the folding calculations are reminiscent of structure prediction problems in protein folding.[26,27] In particular, a novel constrained global optimization problem first introduced for structure prediction of Compstatin using NMR data,[60] and later employed in a generic framework for the structure prediction of proteins, is utilized.[26] The folding formulation represents a general

---

[58] Tobi, D.; Elber, R. *Proteins* **2000**, *41*, 40-46.
[59] Tobi, D.; Shafran, G.; Linial, N.; Elber, R. *Proteins* **2000**, *40*, 71-85.
[60] Klepeis, J. L.; Floudas, C. A.; Morikis, D.; Lambris, J. D. *J. Comp. Chem.* **1999**, *20*, 1354-1370.

nonconvex constrained global optimization problem, a class of problems for which several methods have been developed. In this work, the formulations are solved via the $\alpha\beta\beta$ deterministic global optimization approach, a branch and bound method applicable to the identification of the global minimum of nonlinear optimization problems with twice-differentiable functions.[27,36,37,38,39,60,61]

In addition to identifying the global minimum energy conformation, the global optimization algorithm provides the means for identifying a consistent ensemble of low-energy conformations.[35,61] Such ensembles are useful in deriving quantitative comparisons between the free folding and template-constrained simulations. The relative probability for template stability p_{temp} is calculated by summing the statistical weights for those conformers from the free-folding simulation that resemble the template structure and dividing this sum by the summation of statistical weights over all conformers.

Compstatin is a 13-residue cyclic peptide and a novel synthetic complement inhibitor with the prospect of being a candidate for development as an important therapeutic agent. The binding and inhibition of complement component C3 by Compstatin is significant because C3 plays a fundamental role in the activation of the classical, alternative, and lectin pathways of complement activation. Although complement activation is part of the normal inflammatory response, inappropriate complement activation may cause host cell damage, which is the case in more than 25 pathological conditions, including autoimmune diseases, stroke, heart attack, and burn injuries.[62] The application of the discussed de novo design approach to Compstatin led to the identification of sequences with predicted sevenfold improvements in inhibition activity. These sequences were subsequently experimentally validated for their inhibitory activity using complement inhibition assays.[57]

Summary and Challenge

In the two components of the genomics revolution: (1) sequence to structure, and (2) structure to function, we discussed two significant advances. The first one, the ab initio structure prediction approach ASTRO-FOLD, integrates the two competing points of view in protein folding by employing the thesis of local interactions for the helical formation and the thesis for hydrophobic-hydrophobic residue interactions for the prediction of the topology of β sheets and the location of disulfide bridges. ASTRO-FOLD is based on novel deterministic global optimization and integer linear optimization approaches. The second advance, a novel approach for de novo protein design, introduces in the first stage an explicit mathematical model for the in silico sequence selection that is based on distance-

---

[61] Klepeis, J. L.; Floudas, C. A. *J. Chem. Phys.* **1999**, *110*, 7491-7512.
[62] Sahu, A.; Lambris, J. D. *Immunol. Rev.* **2001**, *180*, 35-48.

dependent force fields and a rigorous treatment of the combinatorial optimization problem, while in the second stage, full atomistic-level folding calculations are introduced to rank the proposed sequences, which are subsequently validated experimentally.

Several important challenges exist; these include new methods for the protein structure prediction that will consistently attain resolutions of about 4-6 Å for all-α, all-β,$\alpha\beta$, and α/β proteins of medium to large size; new approaches that will lead to resolution of protein structures comparable to the existing experimental techniques; novel global optimization methods for sampling in the tertiary structure prediction and refinement; new approaches for the packing of helices in globular and membrane proteins; new computational methods for the structure prediction of membrane proteins; improved methods for protein-protein and protein-DNA interactions; new methods for the determination of active sites and structural and functional motifs; new methods for protein function prediction; new approaches for the design of inhibitors; and systems-based approaches for improved understanding of gene regulatory metabolic pathways and signal transduction networks.

Acknowledgments

The author gratefully acknowledges financial support from the National Science Foundation and the National Institutes of Health (R01 GM52032).

MODELING OF COMPLEX CHEMICAL SYSTEMS RELEVANT TO BIOLOGY AND MATERIALS SCIENCE: PROBLEMS AND PROSPECTS

Richard Friesner
Columbia University

Overview

In this paper, I discuss the future of computational modeling of complex, condensed-phase systems over the next decade, with a focus on biological modeling and specifically structure-based drug design. The discussion is organized as follows. First, I review the key challenges that one faces in carrying out accurate condensed phase modeling, with the analysis centered on the core technologies that form the essential ingredients of a simulation methodology. Next, I examine the use of molecular modeling in structure-based drug design applications. In my presentation, I briefly discussed issues associated with software development, and space limitations do not allow elaboration on that area.

Core Technologies

I focus here on a condensed-phase simulation problem involving the interactions of macromolecular structures, small molecules, and solvent. Periodic solid-state systems have their own special set of difficulties and simplifications, which I do not discuss. The problems of protein structure prediction, protein-ligand binding, and enzymatic catalysis, which are discussed in the next section, fall into this category.

Any condensed-phase simulation protocol that is going to address the above problems requires three fundamental components:

1. *A function describing the energetics of the macromolecule(s) and small molecule(s) as a function of atomic positions.* Such a function can involve direct quantum chemical computation (e.g., via density function theory or MP2 second order perturbation theory methods), a molecular mechanics force field, a mixed QM-MM model, or an empirical scoring function parameterized against experimental data.

2. *A description of the solvent.* If explicit solvent simulation is to be employed, this can simply be another term in the molecular mechanics force field, for example. However, explicit solvent models are computationally expensive and can introduce a substantial amount of noise into the evaluation of relative energies unless extremely long simulation times are used. Therefore, considerable effort has been invested in the development of continuum solvent models, which are relatively inexpensive to evaluate (typically between one and two times the cost of a gas-phase force field evaluation for one configuration) and can achieve reasonable accuracy.

3. *A protocol for sampling configuration space.* If structure prediction is desired, then the objective is to find the minimum free-energy configuration. If one is calculating binding affinities or rate constants, some kind of statistical averaging over accessible phase space configurations is typically required. The total cost of any given calculation is roughly the number of configurations required to converge the phase space average (or to locate the global minimum) multiplied by the computational cost of evaluating the energy (including the contribution of the solvent model if continuum solvation is employed) per configuration. Thus, a key problem is to reduce the number of required configurations by devising more effective sampling methods.

The ability to treat real-world condensed-phase simulation problems, such as structure-based drug design, is dependent on making progress in each of these three areas, to the point where none of them represents an unavoidable bottleneck to achieving adequate accuracy in reasonable CPU times. We discuss each area briefly below, summarizing the current state of the art and prospects for the future.

Energy Models

Quantum Chemical Methods. Quantum chemistry represents the most fundamental approach to computation of the energy of any given atomic configuration. If the electronic Schrödinger equation is solved accurately, the correct answer will be obtained. Since an exact solution is not possible for systems containing more than one electron, the problem here is to solve the Schrödinger equation with approximations that are tractable and yet yield good quantitative accuracy. Following is a summary of the most useful methods that have been developed for doing this:

1. Density function theory provides respectable results for bond energies (2-3 kcal/mol), activation barriers (3-4 kcal/mol), conformational energies (0.5-1.5 kcal/mol), and hydrogen bonding interactions (0.5-1 kcal/mol) with a scaling with system size in the N-N^2 regime (N being the number of electrons in the system). A crucial strength of DFT is its ability to deliver reasonable results across the periodic table with no qualitative increase in computational effort. DFT is currently the method of choice for calculations involving reactive chemistry of large systems, particularly those containing transition metals, and is useful for a wide range of other calculations as well. Systems on the order of 300-500 atoms can be treated using current technology.

2. Second-order perturbation theory (MP2) can provide more accurate conformational energies, hydrogen bond energies, and dispersion energies than DFT (which fails completely for the last of these, at least in current functionals). This makes it the method of choice for computing conformational or intermolecular interactions and developing force fields. The computational scaling is in the N^2 -N^3 range. Systems on the order of 100-200 atoms can be treated using current technology. The use of large basis sets and extrapolation to the basis set limit is necessary if high accuracy is to be achieved for the properties discussed above.

3. Coupled cluster methods (CCSD (T) in particular) provide high-accuracy results (often within 0.1 kcal/mol) for many types of molecules (e.g., organic molecules), but have more difficulties with transition metal-containing species. The method scales as N^7 and at present can conveniently be applied only to small molecules, where it is however quite valuable in producing benchmark results.

Overall, current quantum chemical methods are adequate for many purposes. Improvements in speed and accuracy are ongoing. My personal view is that this does not represent the bottleneck in accurate predictive condensed phase simulations. If one accepts the idea that entirely quantum chemical simulations are not the optimal approach for most problems (and almost certainly not for biological problems), current methods perform well in the context of QM-MM modeling and in producing results for force field development.

Molecular Mechanics Force Fields. There are a number of different philosophies that have been the basis for various force field development efforts. Some groups have relied primarily on fitting parameters to experimental data, while others have introduced a larger component of ab initio quantum chemical data. My view is that extensive use of ab initio data is necessary if one is going to achieve high accuracy and broad coverage of chemical space. Experimental data can be used in conjunction with ab initio data and also used to test the resulting models.

There are two basic issues associated with force field development. The first is the functional form to be used. The simplest cases involve building models for organic molecules that do not consider reactive chemistry. In this case, the standard models (valence terms for stretches, bends, torsions, electrostatic, and van der Waals terms for intermolecular interactions) are sufficient, with the proviso that the electrostatics should be described by higher order multipoles of some sort (as opposed to using only atomic point charges) and that polarization should be introduced explicitly into the model. Simplified models that do not include polarization or higher multipoles appear to yield reasonable structural predictions but may not be able to provide highly precise energetics (e.g., binding affinities, although the improvements attainable with a more accurate electrostatic description have yet to be demonstrated). The second problem is achieving a reliable fit to the quantum chemical and experimental data. This is a very challenging problem that is tedious, labor intensive, and surprisingly difficult technically. While I believe that better solutions will emerge in the next decade, this is an area that definitely could benefit from additional funding.

Current force fields are lacking in both accuracy and coverage of chemical space. Improvements in the past decade have been incremental, although real. With current levels of funding, continued incremental improvement can be expected. Larger-scale efforts (which perhaps will be financed in private industry once people are convinced that bottom line improvements to important problems will result) will be needed to produce improvement beyond an incremental level.

Empirical Models. The use of empirical models is widespread in both academia and industry. There are for example a large number of "statistical" potential functions that have been developed to address problems such as protein folding or protein-ligand docking. The appeal of such models is that they can be designed to be computationally inexpensive and can be fitted directly to experimental data to hopefully yield immediately meaningful results for complex systems. The challenge is that systematic improvement of empirical models beyond a certain point is extremely difficult and becomes more difficult as the basis for the model becomes less physical.

A recent trend has been to combine approaches having molecular mechanics and empirical elements; an example of this is Bill Jorgensen's work on the computation of binding free energies. This is a direction in which much more work needs to be done, much of it of an exploratory variety. In the end, it is unlikely that brute force simulations will advance the solution of practical problems in the next decade (in another 10-20 years, this may actually be a possibility). A combination of physically meaningful functional forms and models with empirical insight gained from directly relevant experiments is more likely to work. Computational chemistry needs to become a full partner with experiment, not try to replace it—the technology to do that simply is not there at this time.

Solvation Models

I concentrate here on continuum models for aqueous solution, although the basic ideas are not very different for other solvents. A great deal of progress has been made in the past decade in developing continuum solvation models based on solution of the Poisson-Boltzmann (PB) equation, as well as approximations to this equation such as the generalized Born (GB) model. These approaches properly treat long-range electrostatic interactions and as such are significantly more accurate than, for example, surface-area-based continuum solvation methods. Progress has also been made in modeling the nonpolar part of the solvation model (i.e., cavity formation, van der Waals interactions) for both small molecules and proteins.

The current situation can be summarized as follows:

• Computational performance of modern PB- and GB-based methods is quite respectable, no more than a factor of 2 more expensive than a gas-phase calculation.

• Accurate results for small molecules can be obtained routinely by fitting experimental data when available.

• The significant issues at this point revolve around the transferability of the parameterization to larger structures. How well do continuum methods describe a small number of waters in a protein cavity, for example, or solvation around a salt bridge (provided that the individual functional groups are well described in isolation)? Our most recent results suggest that there are cases in which a small number of explicit waters is essential and that the transferability problem has not yet been completely solved.

The above problems, while highly nontrivial, are likely to be tractable in the next decade. Thus, my prediction would be that given sufficient resources, robust continuum solvent models (using a small number of explicit waters when necessary) can and will be developed and that this aspect of the model will cease to limit the accuracy. However, a major investment in this area (public or private) is going to be necessary if this is to be accomplished.

Sampling Methods

A wide variety of sampling methods are being developed to address the problems of both global optimization and free-energy calculations. Traditional methods using molecular modeling programs include molecular dynamics, Monte Carlo, conformational searching, and free-energy perturbation techniques. More recently, there have been many new ideas including parallel tempering, replica exchange, quantum annealing, potential smoothing techniques, and so forth. Tests of the newer methods on realistic problems (as opposed to the model problems where they typically are validated initially) have not yet been extensive, so we do not really know which approaches will prove optimal. It does seem likely, however, that significant progress is going to be made beyond that provided by faster processors.

Sampling is a key bottleneck at present in obtaining accurate results in molecular modeling simulations. Obtaining convergence for a complex condensed-phase system is extremely challenging. This is the area in my opinion where prospects are most uncertain and where it is critical to support a lot of new ideas as opposed to just improved engineering of existing approaches. Some advances will come about from faster hardware, but algorithmic improvement should contribute even more if sufficient effort is applied. Until we can converge the sampling, it is going to be very difficult to improve the potential functions and solvation models reliably using experimental input, because there are always questions about whether the errors are due to incomplete sampling as opposed to the model itself.

Structure-based Drug Design

The key problems in structure-based drug design can be enumerated as follows:

1. *Generation of accurate protein structures*, whether in a homology modeling context or simply enumerating low-energy structures given a high-resolution crystal structure. The use of a single protein conformation to assess ligand binding is highly problematic if one wants to evaluate a large library of diverse ligands so as to locate novel scaffolds capable of supporting binding. This is particularly the case for targets such as kinases where there is considerable mobility of loops and side chains in the active site of the protein.

Progress on this problem will come from the ability to rapidly sample possible conformations using an accurate energy functional and continuum solvation model. There is good reason to believe that this can be accomplished in the next three to five years. While perfect accuracy will not be achieved, the generation of structures good enough to dock into and make predictions about binding is a realistic possibility. Experimental data can be used to refine the structures once good initial guesses can be made.

2. *High-throughput docking of ligands into the receptor for lead discovery.* The first objective is to correctly predict the binding mode of the ligand in the receptor. Considerable progress has been made in this direction over the past decade, and continued improvement of the existing technology, as well as use of flexible protein models when required, should yield successful protocols in the next 3-5 years. The second objective is scoring of the candidate ligands—at this stage, one simply wants to discriminate active from inactive compounds. To do this rapidly, an empirical component is required in the scoring. This in turn necessitates the availability of large, reliable datasets of binding affinities. At present the quality and quantity of the publicly available data is problematic; a much greater quantity of data resides in a proprietary form in pharmaceutical and biotechnology companies, but this is inaccessible to most of those working on the development of scoring functions. This problem must be addressed if more robust and accurate scoring functions are to be produced. Given the availability of sufficient data, it seems likely that excellent progress can be made over the next several years.

3. *Accurate calculation of binding free energies for lead optimization.* From a technical point of view, this is the most difficult of the problems that have been posed. At present we do not really know whether the dominant cause of errors in rigorous binding affinity computation should be attributed to the potential functions, solvation model, or sampling; perhaps all three are contributors at a significant level. All one can do is attempt to improve all three components and carry out tests to see whether better agreement with experiment is obtained. My own intuition is that on a 5- to 10-year time scale, there will be some real successes in this type of calculation, but achieving reliability over a wide range of chemistries and receptors is going to be a great challenge.

Overall, there is a good probability that in the next 5-10 years, computational methods will make increasingly important contributions to structure-based drug design and achieve some demonstrable successes.

THE CURRENT STATE OF RESEARCH IN INFORMATION AND COMMUNICATIONS

James R. Heath
University of California, Los Angeles

This paper focuses on an area of nanoelectronics to which chemists have been making substantial contributions in the past five to ten years. It discusses what the chemical challenges are and what the state of the art is in the field of nano-IT.

We have recently prepared a series of aligned semiconductor (silicon) wires, 5 nm in diameter. These are at 16 nm pitch, and 5 mm long; the aspect ratio is 10^7. This means that if a crossbar junction of the silicon were p-doped at a reasonable level, there would be only a 1% chance of the dopant atom being at the wire crossing. Consequently, if the electronics of doped silicon could be brought down to these dimensions, the statistical fluctuations in the density simply mean that classical electronics analysis would fail—one cannot use ohmic assumptions, because one does not know the doping level locally. From a classical computing viewpoint, there are major problems in working at this scale. Each device no longer acts like every other device, leakage currents can dominate in the nanodimension area, and they can cause tremendous parasitic power losses due to very thin gate widths.

One can use molecules to augment the capabilities of the cross wires. It is not difficult to make small structures down to 60 nm, but it is difficult to bring them close together, so density becomes the most challenging synthetic capability.

Work in molecular electronic construction of logic and memory devices has advanced substantially—the Hewlett Packard (HP) group a year ago prepared a 64-bit electronic-based random-access memory. Our group at the University of California, Los Angeles, and the California Institute of Technology has also made a random-access memory, but it is substantially smaller (and fits within a single bit of the work reported by HP). Crossbar structures can still be used for logic. Three-terminal devices can be developed at a single molecule level, and we have done so. I believe that these structures will be the way in which chemists can extrapolate structure-property relationships that go back to synthesis and simple reduced dimensionality circuits.

There are major problems involved in working at this scale—the traditional concepts of Marcus theory may fail because there is no solvent to polarize. Traditional analytical methodology such as NMR, mass spectrometry, and optical techniques are very difficult to use on this length scale. The direct observables are the conductance in two-terminal devices and the gated conductance in three-terminal devices. Using these, we have been able to analyze hysteresis loops and cycling data to indicate that the molecules are stable. Very recent work in our group has shown that one can prepare three-molecule gates and analyze the behavior of these molecular quantum dots that also have the internal structures.

Other concepts become complicated. Is it clear that the molecules behave ohmically in such junctions? How do the energies align? How does gating work within this molecular structure? There are problems with organics, which have mobilities that are many orders of magnitude lower than that of single-crystal silicon. Consequently, there will be a major challenge in making molecules behave like silicon.

The molecular junctions that we have prepared have several advantages: first, they are almost crystal-like, and therefore it seems that they could be chemically assembled. Second, they are quite tolerant of defective components, and are therefore appropriate for the world of chemistry, where reactions never go 100 percent.

Both the HP structure and the structures that we have prepared are really extremely dense: ours are roughly 10^{12} bits per centimeter. The volume density of information storage in the brain is roughly 10^{12} per cubic centimeter, so the density of these molecular structures is extremely high.

Finally, it is important to think about architectures. Molecular devices have now been demonstrated, and the fabrication of several molecular devices has been clarified over the past five years. However, architecture is more complicated, and it requires the fabrication, alignment, interaction, and behavior of many devices. Key questions are now being addressed by scientists in places such as Carnegie Mellon, Stanford, HP, and Caltech: How big should the memory be? How big should the wires be? How much should be devoted to routing? And how much gain needs to be put in?

Some of the structures that I showed you of crossbars had 2200 junctions; they were all made; there weren't any broken components. So at the level of 10^4 bits, it seems possible to do high-fidelity fabrication. This is the start of a true molecular electronics.

MULTISCALE MODELING

Dimitrios Maroudas
University of Massachusetts

In the next decade, multiscale modeling will be a very important area of chemical science and technology—in terms of both needs and opportunities. The field has emerged during the past 10 years in response to the need for an integrated computational approach toward predictive modeling of systems that are both complex and complicated in their theoretical description. The applications to date have involved complexity in physical, chemical, or biological phenomena or have addressed problems of material structure and function. The intrinsic characteristics of all such systems are the multiplicity of length scales due to, e.g., multiple structural features, multiplicity of time scales due to multiple kinetic phenomena that govern processing or function, the strong nonlinear response as these systems operate away from equilibrium, and the large number or broad range of operating parameters that are involved in practical engineering applications.

From the viewpoint of theory and computation, the major challenge in this area will be to establish rigorous links between widely different theoretical formalisms—quantum mechanics, classical statistical mechanics, continuum mechanics, and so on—that span a very broad range of space and time scales and are used to explore broad regions of parameter space. The practical goal will be to derive, as rigorously as possible, relationships between processes, structure, function, and reliability, and to use them to develop optimal engineering strategies.

The core capabilities of multiscale modeling consist of computational methods that have been developed over many decades and are now used to compute properties and model phenomena. Figure 1 illustrates some of these methods in a schematic diagram of length versus time: *Computational Quantum Mechanics* for accurate calculation of properties and investigation of small and fast phenomena, *Statistical Mechanics* for semi-empirical modeling for mechanistic understanding and, at the much larger scale, *Continuum Mechanics* for macroscopic modeling. Between these last two is the *Mesoscopic-Microstructural Scale*, which has been an important motivator for the development of multiscale modeling techniques in the area of materials science. Ultimately, what one would like to do, from an engineering viewpoint at least, is use all these methodologies to explore vast regions of parameter space, identify critical phenomena, promote critical phenomena that improve the behavior of a system, and avoid critical phenomena that lead to failure.

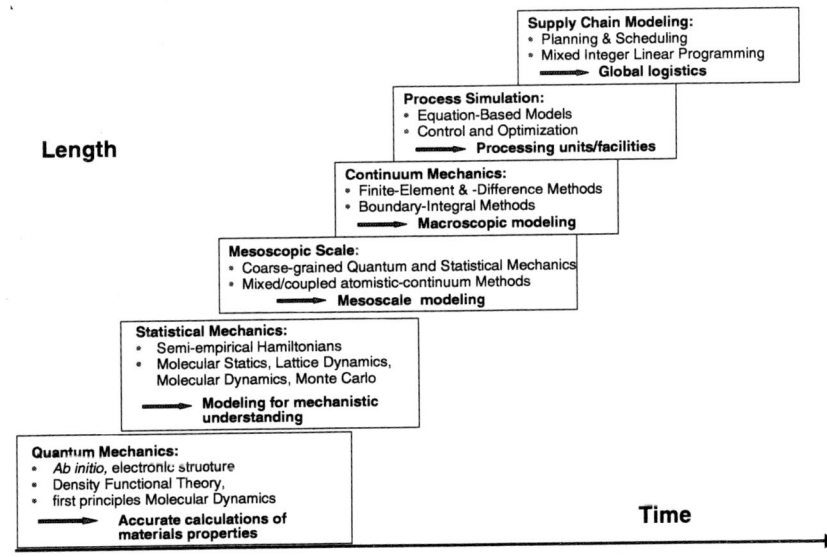

FIGURE 1 Modeling elements and core capabilities.

Strict requirements must be imposed if multiscale modeling is to become a powerful predictive tool. In particular, we need to deal with issues of *accuracy* and *transferability* through connection to first principles because phenomenological models are not transferable between widely different environments. The question is unavoidable: Is system level analysis—starting from first principles—feasible for complex systems? In certain cases, the realistic answer would be an immediate "no." In general, the optimistic answer is "not yet," but the stage has been set for tremendous progress in this direction over the next decade. In addition to developing novel, robust multiscale computational methods, *fundamental mechanistic understanding* will be invaluable in order to enable computationally efficient schemes and to steer parametric studies and design of experiments.

I suggest classifying the approaches for multiscale modeling into two categories:

1. *Serial Strategies:* Different-scale techniques are implemented sequentially in different computational domains at different levels of discretization.

2. *Parallel Strategies:* Different-scale techniques are implemented simultaneously in the same computational domain that is decomposed appropriately.

A number of significant trends are noteworthy:

• A variety of multi-space-scale techniques are emerging from efforts on different applications where phenomena at the mesoscopic scale are important.

• Efforts to push the limits of the core capabilities continue as illustrated by molecular-dynamics (MD) simulations with hundreds of millions of atoms.

• Kinetic Monte Carlo (KMC) methods are used increasingly to extend the time-scale limitation of atomistic simulations. These, in turn, are driving the creation of methodology for properly treating structural complexity in KMC schemes and for assessing the completeness and accuracy of the required database, i.e., determining whether all of the important kinetic phenomena are included and accurately calculating transition probabilities.

• Methods are being developed for accelerating the dynamics of rare events—either by taking advantage of physical insights about the nature of transition paths or by using numerical methods to perform atomistic simulations for short periods and then project forward over large time steps.

• Multiscale models are making possible both the integration of insight between scales to improve overall understanding and the integration of simulations with experimental data. For example, in the case of plastic deformation of metals, one can incorporate constitutive theory for plastic displacements into macroscopic evolution equations, where parameterization of the constitutive equations is derived from analysis of MD simulations.

• Methods are being explored for enabling microscopic simulators to perform system-level analysis—mainly numerical bifurcation and stability analysis—to predict and characterize system instabilities and effectively compute the stability domains of the system. The central question is, can we predict the onset of critical phenomena that lead to phase, structural, or flow transitions? Such transitions may lead to function improvement (e.g., prediction of disorder-to-order transitions, or failure).

In conclusion, over the past decade, various multiscale methods and models have been developed to couple widely different length scales, accelerate rare-event dynamics, and explore the parametric behavior of complex systems. These multiscale methods have been applied successfully to various problems in physical chemistry, chemical engineering, and materials science. Over the next decade, these and other new methods will enable truly predictive analyses of complex chemical, material, and biological systems. These will provide powerful tools for fundamental understanding, as well as technological innovation and engineering applications. The development of multiscale methods will generate tremendous research opportunities for the chemical sciences, and the integration of multiscale methods with advances in software and high-performance computing will be strategically important. The new opportunities also will present an educational challenge—to enable students, researchers, and practitioners to understand deeply what is going on throughout the physical scales and parameter space so they can develop intuition and understanding of how best to carry out simulations of complex systems.

THE COMING OF AGE OF COMPUTATIONAL SCIENCES

Linda R. Petzold
University of California, Santa Barbara

The workshop program overstated the title of my contribution. Rather than "The Coming Age of Computational Sciences," I focus on this discipline's coming *of* age. By this, I mean that computational science is like a teenager who has a large potential for rapid growth and a great future but is in the midst of very rapid change. It is like a love affair with the sciences and engineering that is really going to be great.

I explain how I see computational science—what it is, where it is going, and what is driving the fundamentals that I believe are taking place, and finally, how all this relates to you and to the chemical sciences.

I start with some controversial views. Many of us see computation as the third mode of science, together with theory and experiment, that is just coming of age. What could be driving that? Other presentations at this workshop have noted the tremendous growth of the capabilities of both hardware and algorithms. Then I discuss what I see as the emerging discipline of computational science and engineering—or maybe it's the antidiscipline, depending on how one looks at it. I hope that discipline and antidiscipline don't interact in the same way as matter and antimatter!

Next I turn to the revolution in engineering and the sciences at the micro scale, what this means for computational science, the implications for the kinds of tools we are going to need, and what is already happening in response to the rapidly increasing application of engineering to biology. Finally, I address some current trends and future directions, one of which is going beyond simulation. In many cases, simulation has been successful, but now people are getting more ambitious. As algorithms and software get better and the computer gets faster, we want to do more—and this will be a big theme in computational science. What will be the role of computer science in the future of computational science?

Figure 1 illustrates computational speedup over the last 30 years. The lower

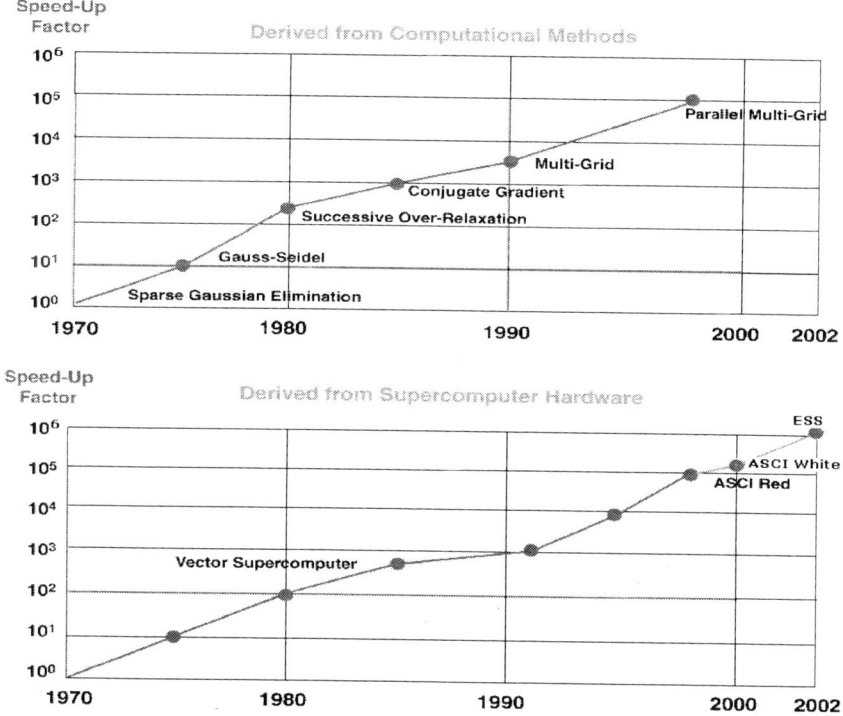

FIGURE 1 Speedup resulting from software and hardware developments. (Updated from charts in *Grand Challenges: High Performance Computing and Communications,* Executive Office of the President, Office of Science and Technology Policy Committee on Physical, Mathematical and Engineering Sciences, 1992; SIAM Working Group on CSE Education, *SIAM Rev.,* **2001**, *43:1,* 163-177).

plot, which shows speedup (relative to computer power in 1970) derived from supercomputer hardware, is just another restatement of Moore's Law.[1] The three last data points are ASCI Red and ASCI White (the big DOE machines) and the famous Earth Simulator System (ESS).

The ESS is right on schedule for the computer power that we would expect based on the trends of the last 30 years. The architecture is vector parallel, so it's a combination of the kind of vectorization used 20 years ago on Cray supercomputers, with the parallelization that has become common on distributed machines over the last decade. It's no surprise that this architecture leads to a fast computer. Probably the main reason that it's newsworthy is that it was built in Japan rather than the United States. Politicians may be concerned about that, but as scientists I think we should just stay our course and do what we feel is right for science and engineering.

The upper graph in Figure 1 is my favorite because it is directly related to my own work on algorithms. It shows performance enhancement derived from computational methods, and it is based on algorithms for solving a special problem at the heart of scientific computation—the solution of linear systems. The dates may seem strange—for example, Gaussian elimination came earlier than 1970—but they illustrate when these algorithms were introduced into production-level codes at the DOE labs. Once again, the trend follows Moore's Law.

My view on what these numbers mean to us, and why computational science and engineering is coming of age, relates to the massive increases in both computer and algorithm power. In many areas of science and engineering, the boundary has been crossed where simulation, or simulation in combination with experiment, is more effective in some combination of time, cost, and accuracy, than experiment alone for real needs. In addition, simulation is now a key technology in industry. At a recent conference, I was astonished to see the number of companies using computer applications to address needs in chemical processing. There is also a famous example in my field, the design of the Boeing 777—an incredibly complex piece of machinery—in which simulations played a major role.

The emerging discipline of computational science and engineering is closely related to applied mathematics (Figure 2). There were early arguments that it shouldn't be called mathematics but applied mathematics—illustrating the disciplinary sensitivities of the fields—but computer science and engineering also overlaps strongly with science, engineering, and computer science. It lies in the twilight zone between disciplines. This is a great area of opportunity, because there is a lot of room there for growth!

---

[1] Moore originally stated that "The complexity for minimum component costs has increased at a rate of roughly a factor of two per year," Moore, G. E., *Electronics* **1965,** *38 (8)* 114-17; This has been restated as "Moore's Law, the doubling of transistors every couple of years," *(http://www.intel.com/ research/silicon/mooreslaw.htm).*

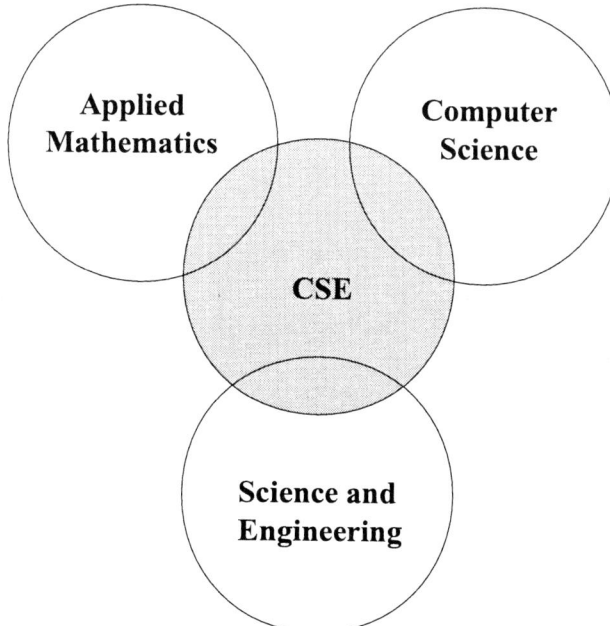

FIGURE 2 Computer Science and Engineering (CSE) focuses on the integration of knowledge and methodologies from computer science, applied mathematics, and engineering and science.

In an academic setting, it's easy to become mesmerized by disciplinary boundaries. We identify with those in our own discipline, but here I'm just looking from the outside. Some have suggested that the structure is like a religion, so your department would be like a branch of the church of science. Consequently, it feels a little bit heretical to suggest that there is a lot of value in integrating the disciplines. Nevertheless, I believe there is a lot of science and technology to be done somewhere in that murky area between the disciplines.

The major development that is driving change, at least in my world, is the revolution at the micro scale. Many people are working in this area now, and many think it will be as big as the computer revolution was. Of particular importance, the behavior of fluid flow near the walls and the boundaries becomes critical in such small devices, many of which are built for biological applications. We have large molecules moving through small spaces, which amounts to moving discrete molecules through devices. The models will often be discrete or stochastic, rather than continuous and deterministic—a fundamental change in the kind of mathematics and the kind of software that must be developed to handle these problems.

For the engineering side, interaction with the macro scale world is always going to be important, and it will drive the multiscale issues. Important phenomena occurring at the micro scale determine the behavior of devices, but at the same time, we have to find a way to interact with those devices in a meaningful way. All this must happen in a simulation.

Many multiscale methods have been developed across different disciplines. Consequently, much needs to be done in the fundamental theory of multiscale numerical methods that applies across these disciplines. One method is famous in structural materials problems: the quasi-continuum method of Tadmor, Ortiz, and Philips.[2] It links the atomistic and continuum models through the finite element method by doing a separate atomist structural relaxation calculation on each cell of the finite element method mesh, rather than using empirical constitutive information. Thus, it directly and dynamically incorporates atomistic-scale information into the deterministic scale finite element method. It has been used mainly to predict observed mechanical properties of materials on the basis of their constituent defects.

Another approach is the hybrid finite element-molecular dynamics-quantum mechanics method, attributed to Abraham, Broughton, Bernstein, and Kaxiras.[3] This method is attractive because it is massively parallel, but it's designed for systems that involve a central defective region, surrounded by a region that is only slightly perturbed from the equilibrium. Therefore it has limitations in the systems that it can address. A related hybrid approach was developed by Nakano, Kalia, and Vashista.[4] In the totally different area of fluid flow, people have been thinking about these same things. There has been adaptive mesh and algorithm refinement, which in the continuum world has been very successful. It's a highly adaptive way of refining the mesh that has been useful in fluid dynamics. These researchers have embedded a particle method within a continuum method at the finest level (using an adaptive method to define the parts of the flow that must be refined at the smaller scale) and applied this to compressible fluid flow.

Finally, one can even go beyond simulation. For example, Ioannis Kevrekidis[5] has developed an approach to computing stability and bifurcation analysis using time steppers, in which the necessary functions are obtained directly from atomistic-scale simulations as the overall calculation proceeds. This

---

[2]Shenoy, V. B.; Miller, R.; Tadmor, E. B.; Phillips, R.; Ortiz M. *Physical Review Letters* **1998**, *80*, 742-745; Miller, R.; Tadmor, E. B., *Journal of Computer-Aided Materials Design* **2003**, in press.
[3]Abraham, F. F.; Broughton, J. Q.; Bernstein, N.; Kaxiras, E. *Computers in Physics* **1998**, *12*, 538-546.
[4]Nakano, A.; Kalia, R. K.; Vashishta, P. *VLSI Design* **1998**, *8*, 123; Nakano, A.; Kalia, R. K.; Vashishta, P.; Campbell, T. J.; Ogata, S.; Shimojo, F.; Saini, S. *Proceedings of Supercomputing 2001* (http://www.sc2001.org).
[5]Kevrekidis, Y. G.; Theodoropoulos, K.; Qian, Y.-H. *Proc. Natl. Acad. Sci. U.S.A.* **2000**, *97*, 9840.

is difficult because it must take into account the boundary conditions on the small-scale simulations. There are so many of these multiscale problems and algorithms that the new *Multiscale Modeling and Simulation Journal* is devoted entirely to multiscale issues.

Another big development is taking place where engineering meets biology (Box 1). These two different worlds will vastly augment each other, but what does it mean for computation? A huge number of multiscale problems exist along with problems that involve understanding and controlling highly nonlinear network behavior. At a recent meeting in the area of systems biology, someone suggested that it would require 140 pages to draw a diagram for the network behavior of *E. coli*. A major issue is uncertainty. The rate constants are unknown, and frequently the network structure itself is not well understood. We will need to learn how to deal with uncertainty in the network structure and make predictions about what it means. It would be useful to be able to tell experimentalists that we think something is missing in this structure.

BOX 1
Engineering Meets Biology: Computational Challenges

- Multiscale simulation
- Understanding and controlling highly nonlinear network behavior (140 pages to draw a diagram for network behavior of *E. coli*)
- Uncertainty in network structure
- Large amounts of uncertain and heterogeneous data
- Identification of feedback behavior
- Simulation, analysis, and control of hybrid systems
- Experimental design

We have large amounts of both uncertain and heterogeneous data. Biologists are obtaining data any way they can and accumulating the data in different forms—discrete data, continuous data, old data, and new data. We must find a way to integrate all these data and use them as input for the computations.

A key challenge in systems biology is identification of the control behavior. Biology is an engineered system—it just was engineered by nature and not by us. However, we would like to understand it and one way to do that is to treat it as an engineered system. It is a system that has many feedback loops, or else it wouldn't be nearly as stable as it is. We need to identify the feedback behavior, which is difficult using only raw simulations and raw data. Again, there are many variables arising from different models, leading to something that control engineers call a hybrid system. Such a system may have continuous variables, Boolean variables, and Bayesian variables. Until now, computational scientists haven't dealt with such systems, but we must learn to do so quickly.

Experimental design offers a unique opportunity for computation to contribute in a huge way. Often it is not possible, or nobody has conceived of a way, to do experiments in biology that can isolate the variables, because it's difficult to do in vivo experiments—so the data come out of murkier experiments. It would be of great value to the experimental world if we could design computations that would allow us to say, "Here is the kind of experiment that could yield the most information about the system."

An important multiscale problem arises from the chemical kinetics within biological systems, specifically for intracellular biochemical networks. An example is the heat shock response in *E. coli*, for which an estimated 20 to 30 sigma-32 molecules per cell play a key role in sensing, in the folding state of the cell, and in regulating the production of heat shock proteins. The method of choice is the stochastic simulation algorithm, in which molecules meet and react on a probabilistic basis, but the problem becomes too large because one also must consider the interactions of many other molecules in the system. It is a challenge to carry out such simulations in anything close to a reasonable time, even on the largest computers.

Two important challenges exist for multiscale systems. The first is multiple time scales, a problem that is familiar in chemical engineering where it is called stiffness, and we have good solutions to it. In the stochastic world there doesn't seem to be much knowledge of this phenomenon, but I believe that we recently have found a solution to this problem. The second challenge—one that is even more difficult—arises when an exceedingly large number of molecules must be accounted for in stochastic simulation. I think the solution will be multiscale simulation. We will need to treat some reactions at a deterministic scale, maybe even with differential equations, and treat other reactions by a discrete stochastic method. This is not an easy task in a simulation.

I believe that some trends are developing (Box 2), and in many fields, we are moving beyond simulation. As a developer of numerical software, in the past 10 years, I've seen changes in the nature of computational problems in engineering. Investigators are asking questions about such things as sensitivity analysis, optimum control, and design optimization. Algorithms and computer power have reached a stage where they could provide success, but we are ambitious, and want to do more. Why do people carry out simulations? Because they want to design something, or they want to understand it.

I think the first step in this is sensitivity analysis, for which there has been a lot of work in past decade. The forward method of sensitivity analysis for differential equations has been thoroughly investigated, and software is available. The adjoint method is a much more powerful method of sensitivity analysis for some problems, but it is more difficult to implement, although good progress has been made. Next it moved to PDEs, where progress is being made for boundary conditions and adaptive meshes.

> **BOX 2**
> **Beyond Simulation: Computational Analysis**
>
> *Sensitivity Analysis*
> - Forward method—ODE-DAE methods and software; hybrid systems
> - Adjoint method—ODE-DAE methods and software
> - Forward and adjoint methods—PDE methods with adaptive mesh refinement
> - Multiscale, stochastic—still to come
>
> *Uncertainty Analysis*
> - Polynomial chaos, deterministic systems with uncertain coefficients; still to come: multiscale
> - Stochastic
> - Many other ideas
>
> *Design Optimization–Optimal Control*

Uncertainty analysis was mentioned many times at the workshop, and still to come are multiscale systems, fully stochastic systems, and so on. And then we will move toward design optimization and optimal control. There is a need for more and better software tools in this area. Even further in the future will be computational design of experiments, with the challenge of learning the extent to which one can learn something from incomplete information. Where should the experiment be done? Where does the most predictive power exist in experiment space and physical space? Right now, these questions are commonly answered by intuition, and some experimentalists are tremendously good at it, but computations will be able to help. Some examples are shown in Box 3.

I think computer science will play a much larger role in computational science and engineering in the future. First, there are some pragmatic reasons. All of the sciences and engineering can obtain much more significant help from software tools and technology than they have been receiving. A really nice example is source code generation—codes that write the codes. A very successful application is that of automatic differentiation codes developed at Argonne National Laboratory. These codes take input functions written in FORTRAN or C, while you specify where you have put your parameters and instruct them to differentiate the function with respect to those parameters. In the past, this has been a big headache because we had to do finite differences of the function with respect to perturbations of the parameter. In principle this a trivial thing—first-semester calculus—but it's actually very difficult to decide on the increment for the finite

> **BOX 3
> More Computational Analysis**
>
> - *Design of experiments:* to what extent can one learn something from incomplete information? Where is the most predictive power?
>
> - *Determination of nonlinear structure:* multiscale, stochastic, hybrid, Bayesian, Boolean systems
>   - Bifurcation
>   - Mixing
>   - Long-time behavior
>   - Invariant manifolds
>   - Chaos
>   - Control mechanisms: identifying feedback mechanisms
>
> - *Reduced and simplified models:* deterministic, multiscale, stochastic, hybrid systems, identify the underlying structure and mechanism
>
> - *Data analysis:* revealing the interconnectedness, dealing with complications due to data uncertainties

difference. There is a very tenuous trade-off between round-off and truncation error, for which there sometimes is no solution—particularly for chemistry problems, which tend to be badly scaled. The new approach writes the derivative function in FORTRAN or C, and the result is usually perfect, eliminating all of the finite differencing errors. Moreover, it doesn't suffer from the explosion of terms that were encountered with symbolic methods. The methodology uses compiler technology, and you might be surprised at what the compiler knows about your program. It's almost a Big Brother sort of thing. This has enabled us to write much more robust software, and it has greatly reduced the time needed for debugging.

Another thing we can do use source-code generation to fix the foolish mistakes that we tend to make. For example, when I consult with companies, I often hear the problem of simulation software that has worked for a long time but suddenly failed when something new was introduced—and the failure appears to be random. I always ask the obvious question, "Did you put an 'if' statement in your function?" And they always say, "No, I know about that potential problem: I wouldn't be stupid enough to do that." But after many hours of looking at multiple levels of code (often written by several generations of employees who have all gone on to something else), we find that "if" statement.

In any case, what Paul Barton has done to source code generation is to go in and just automatically fix these things in the codes, so that we don't have to go in and do it. And sometimes it's a really hard thing to go into some of these codes, dusty decks, or whatever, written by people who have departed long ago, so this is really making things easier.

Another big issue is that of user interfaces for scientific software. User interfaces generate many complaints, and with good reason! I am embarrassed to admit that my own codes and the scientific engineering software with which I come in contact regularly are in the dinosaur age. If our interfaces were better—up to the standards of Microsoft Office, for example—we would have an easier time attracting students into scientific and engineering fields. It certainly can't excite them to deal with those ugly pieces of 30-year-old codes. On the other hand, we aren't going to rewrite those codes, because that would be tremendously tedious and expensive.

There have been some good developments in user interface technology for scientific computing, and some exceptions to the sad state of most of our software interfaces. In fact, I think that the first one is obvious: MATLAB.[6] It is an excellent example of the value of a beautiful interface, but it has been directed primarily at relatively small problems.

Computer science technology can be used to enable the semiautomatic generation of graphical user interfaces. This is also being done at Sandia National Laboratories and at Lawrence Livermore National Laboratory with the MAUI code. There are ugly codes in chemistry. We all know it. And there are many dusty decks out there in industry and national laboratories. We'd like to just keep those codes but use compiler technology to bring up the kind of interfaces for the user. A collaboration with those users will be necessary, which is why the method is semiautomatic; but it can become a reality without every user needing to learn the latest user-interface technology.

Finally, my major point is that computer science will play a much larger role. There is a deep reason for that. It isn't just machines, compilers, and source-code generation that will help you with your research. Those will be nice and undoubtedly will be useful, but they are not the main issue.

At the smaller scales—and this is the fundamental change—we are dealing with and manipulating large amounts of discrete stochastic, Bayesian, and Boolean information. The key word is information. In the past, we manipulated continuum descriptions, which are averages. Those were nice, but now we must manipulate discrete data—heterogeneous data. Who has been thinking about that for the last 20 or 30 years? These problems form the core of computer science.

Living simultaneously in both a computer science department and a mechanical engineering department, I know that the communications between engineering and computer science have not always been as good as they could be. But we are all researchers, and the needs of science will make the communication happen.

---

[6]Originally developed as a "matrix laboratory" to provide access to matrix software, MATLAB integrates mathematical computing, visualization, and a programming language for scientific and technical computing: *http://www.mathworks.com/*

SIMULATION IN MATERIALS SCIENCE

George C. Schatz
Northwestern University

Introduction

The primary goal of this paper is to provide examples of problems in nanomaterials where computational methods are able to play a vital role in the discovery process. Three areas of research are considered, and in each of these areas, an important scientific problem is identified. Then we consider the current computational and theoretical tools that one can bring to bear on the problem and present representative results of applications to show how computation and theory have proven to be important. The three areas are (1) optical properties of gold nanoparticle aggregates, (2) melting of DNA in the gold nanoparticle aggregates, and (3) oxygen atom erosion of polymers in low earth orbit conditions. The three theories and computational methods being highlighted are (1) computational electrodynamics, (2) empirical potential molecular dynamics, and (3) direct dynamics–Car-Parinello molecular dynamics.

Optical Properties of Gold Nanoparticle Aggregates

Although colloidal gold has been known for thousands of years as the red color of stained glass windows, only very recently has it found a use in the medical diagnostics industry as a sensor for DNA. In this work by Chad Mirkin and colleagues at Northwestern, single-stranded oligonucleotides are chemically attached to colloidal gold particles, typically 10 nm in diameter. If an oligonucletide (that is from DNA that one wishes to detect and is complementary to the single stranded portions attached to the gold particles) is added to the system, DNA hybridization results in aggregation of the particles, inducing a color change from red to blue that signifies the presence of the complementary oligonucleotide. Small-angle x-ray scattering studies of these aggregates indicate that although the aggregates are amorphous, the nearest-neighbor spacing between particles is approximately equal to the length of the linker oligonucleotide.

The challenge to theory in this case is to develop an understanding of the optical properties of the aggregates so that one can determine the optimum particle size and optimum linker length, as well as the variation in optical properties that can be achieved using other kinds of nanoparticles. The basic theory needed for this problem is classical electrodynamic theory because this is known to describe the optical properties of metal particles accurately, provided only that we have the dielectric constants of the particles and surrounding medium and that we have at least a rough model for the aggregate structure.

Computational electrodynamics has made great progress during the past decade, with the result that codes are now available for describing the optical properties of assemblies of nanoparticles with upwards of tens of thousands of particles. However there are many choices for methods and codes, including a variety of finite element (Grid-based) methods (discrete dipole approximation, finite difference time domain, multiple multipoles, dyadic Green's function) and also a variety of coupled particle methods, of which the most powerful is the T-matrix method but which in the present case can be simplified to coupled dipoles for most applications. The codes for this application are largely public domain, but important functionality can be missing, and we have found it necessary to make revisions to all the codes that we have used.

Fortunately the nanoparticle aggregate electrodynamics problem is relatively easy to solve, and many of the codes that we examined give useful results that not only explain the observed color change, but tell us how the results vary with particle size, DNA length, and optical constants. Figure 1 shows a typical comparison of theory and experiment. We learned enough from our studies of this problem that we were able to develop a simple, almost analytical theory based on a dynamic effective medium approximation. In addition, the methods that we considered have been useful for studying more challenging electrodynamics problems that involve ordered assemblies of nonspherical particles in complex dielectric environments.

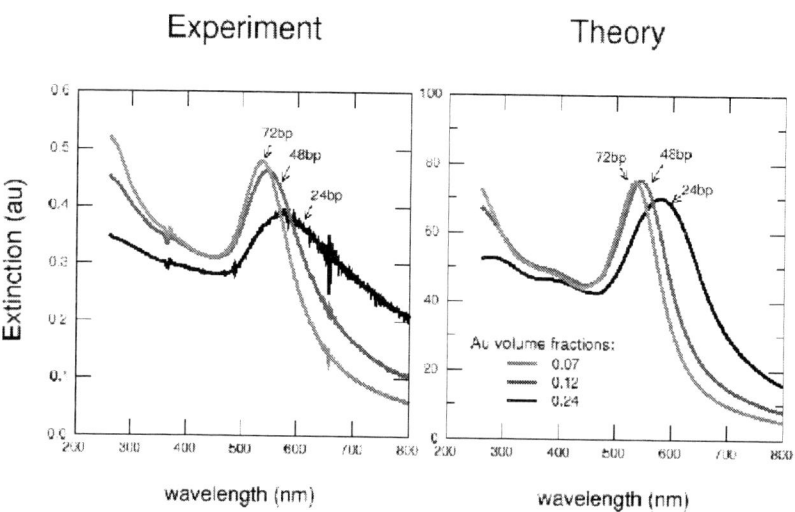

FIGURE 1 Absorption spectra of DNA-linked gold nanoparticle aggregates for 24, 28 and 72 base-pair DNA linkers, comparing theory (light scattering calculations, right) and experiment (left).

Melting of DNA That Links Nanoparticles

The DNA-linked gold nanoparticle aggregates just described have another use for DNA detection that is both very important and distinctly nanoscale in nature. When the aggregates are heated, it is found that they "melt" (i.e., the double helix unravels into single-stranded oligonucleotides) at about 50 °C over a narrow temperature range of about 3 °C. This melting range is to be contrasted with what is observed for the melting of DNA in solution, which typically has a 20°C range for the same length oligonuclides. This is important to DNA testing, as single base pair mismatches, insertions, and deletions can still lead to aggregate formation, and the only way to identify these noncomplementary forms of DNA is to subject the aggregates to a thermal wash that melts out the noncomplementary linkers but not the complementary ones. The resolution of this process is extremely sensitive to the melting width. In view of this, it is important to establish the mechanism for the narrow melting transition.

Recently we have developed a cooperative melting model of DNA melting for DNA-linked gold nanoparticle aggregates. This model considers the equilibrium between different partially melted aggregates, in which each stage involves the melting of a linker oligonucleotide, but there are no other structural changes (at least initially). In this case, the equilibria can be written

$$D_N = D^N_{-1} + Q + nS$$
$$\ldots \quad \ldots$$
$$D_2 = D_1 + Q + nS$$
$$D_1 = D_0 + Q + nS$$

where D_N is the aggregate with N linkers, Q is the linker, and S stands for free counterions that are released with each linker. The key to the cooperative mechanism is to realize that each successive step becomes easier due to the loss of counterions from the aggregate. Here, we are guessing that when the DNAs are sufficiently close that their ion clouds overlap, the ion clouds mutually stabilize the hybridized state (due to improved screening of the phosphate interactions). With this assumption, the equilibrium collapses to a single expression:

$$DN = D_0 + NQ + nNS$$

in which N targets and their complement of counterions come out cooperatively. Equilibrium concentrations based on this expression are consistent with a variety of measurements on the gold nanoparticle aggregates. Figure 2 shows the sharpening of the melting curve that occurs as the number of DNA linkers per nanoparticle is increased, while Figure 3 shows a fit to experimental data using this model.

While this works, the key assumption of the theory, that the ion clouds over-

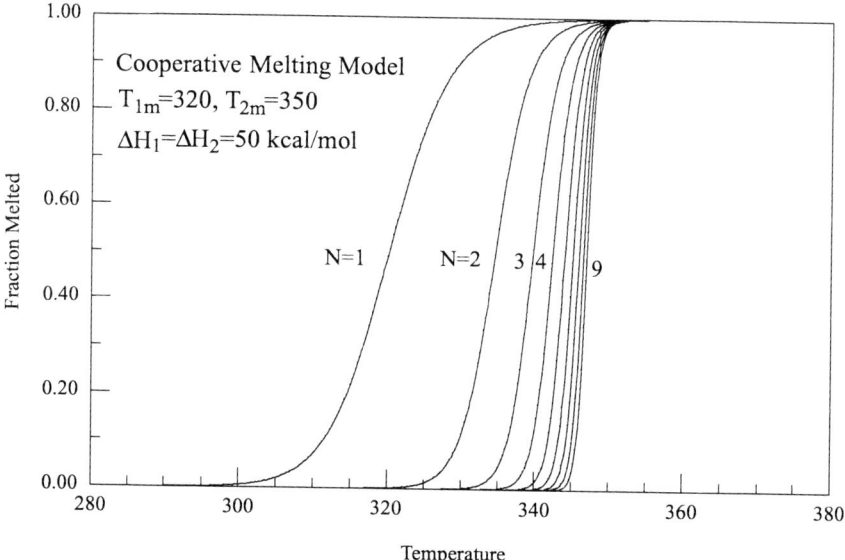

FIGURE 2 Cooperative melting mechanism: dependence of melting curves on number of DNA linkers per nanoparticle.

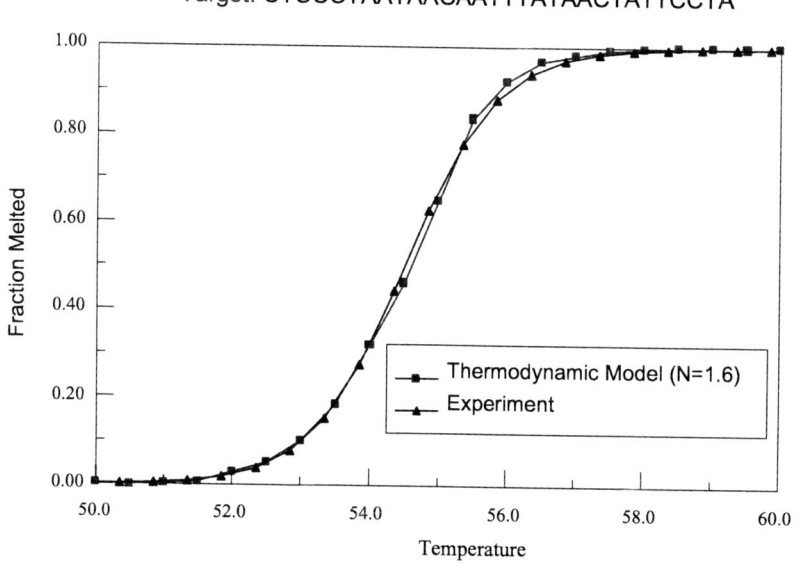

FIGURE 3 Cooperative melting mechanism: fit to experiment.

lap and the DNA is stabilized needs to be proven, and crucial information about the range of the cooperative mechanism needs to be determined. To do this, we must simulate the ion atmosphere around DNA dimers, trimers and other structures, ideally with gold particles also nearby. This could probably be done with an electrostatic model, but there is concern about the validity of such models for singly charged counterions. To circumvent this, we have considered the use of atomistic molecular dynamics with empirical force fields. Such methods have already been used to determine charge cloud information concerning a single duplex, but the calculations had not been considered for aggregates of duplexes.

Fortunately the methodology for simulating DNA and its surrounding charge cloud is well established based on the Amber suite of programs. This suite allows for the inclusion of explicit water and ions into the calculation, as well as particle-mesh Ewald sums to take care of long-range forces. Previous work with Amber, explicit ions, and Ewald summation has demonstrated that the ion distribution around duplex DNA is consistent with available observations in terms of the number of ions that are condensed onto the DNA, and the extent of the counterion cloud around the DNA. Given this, it is therefore reasonable to use the same technology to examine the counterion distribution around pairs (and larger aggregates) of DNA duplexes and how this varies with bulk salt concentration.

My group has now performed these simulations, and from this we have established the relationship between bulk salt concentration and the counterion concentration close to the DNA. By combining these results with the measured variation of DNA melting temperature with bulk salt concentration, we have determined that the decrement in melting temperature associated with loss of a DNA target from an aggregate is several degrees, which is sufficiently large that cooperative melting can occur. In addition we find that the range of interaction of the counterion atmospheres is about 4 nm, which establishes the DNA density needed to produce cooperative melting.

Oxygen Atom Erosion of Polymers in Low Earth Orbit Conditions

The exterior surfaces of spacecraft and satellites in low earth orbit (LEO) conditions are exposed to a very harsh environment. The most abundant species is atomic oxygen, with a concentration of roughly 10^9 atoms per cm^3. When one factors in the spacecraft velocity, one finds that there is one collision per second per exposed surface site on the spacecraft, and the oxygen atoms typically have 5 eV of energy, which is enough to break bonds. When this is combined with other LEO erosion mechanisms that involve ions, electrons, UV photons, and other high-energy neutrals, one finds that microns of exposed surface can be eroded per day unless protective countermeasures are imposed. This problem has been known since the beginning of space flight, but the actual mechanism of the erosion process has not been established. In an attempt to rectify this situation, and perhaps to stimulate the development of new classes of erosion resistant materials, I have been collaborating with a team of AFOSR funded faculty—Steven Sibener, Luping Yu, Tim Minton, Dennis Jacobs, Bill Hase, Barbara Garrison, John Tully—to develop theory to model the erosion of polymers and other materials, and to test this theory in conjunction with laboratory and satellite experiments. Of course this research will also have spin-offs to other polymer erosion processes important in the transportation and electronics industries.

We are only at the beginning of this project at the moment, so all of our attention thus far has been directed at understanding the initial steps associated with the impact of high-velocity oxygen atoms with hydrocarbon polymer surfaces. Even for this relatively simple class of materials, there has been much confusion as to the initial steps, with the prevailing wisdom being that $O(^3P)$, the ground state of oxygen, can only abstract hydrogen atoms from a hydrocarbon, which means that in order to add oxygen to the surface (as a first step to the formation of CO and CO_2), it is necessary for intersystem crossing to switch from triplet to singlet states, thereby allowing the oxygen atom to undergo insertion reactions. This wisdom is based on analogy with hydrocarbon combustion, where hydrogen abstraction reactions are the only mechanism observed when oxygen interacts with a saturated hydrocarbon.

In order to explore the issue of oxygen-atom reactions with hydrocarbons in an unbiased way, we decided to use "direct dynamics" (DD), i.e., molecular dynamics simulations in which the forces for the MD calculations are derived from electronic structure calculations that are done on the fly. Although this idea has been around for a long time, it has become practical only in the last few years as a result of advances in computational algorithms for electronic structure calculations and the advent of cheap distributed parallel computers. We have studied two approaches to this problem: the Car-Parinello molecular dynamics (CPMD) method using plane wave basis functions and ultrasoft pseudopotentials, and conventional Born-Oppenheimer molecular dynamics (BOMD) using Gaussian orbitals. Both calculations use density functional theory, which is only capable of 5 kcal/mol accuracy for stationary point energies, but should be adequate for the high energies of interest to this work. Both methods give similar results when applied to the same problem, but CPMD is more useful for atom-periodic surface dynamics, while BOMD is better for cluster models of surfaces or disordered surfaces.

Our calculations indicate that there are important reaction pathways that arise with 5-eV oxygen atoms that have not previously been considered. In particular we find that oxy radical formation is as important as hydrogen atom abstraction and that C-C bond breakage also plays a role. Both of these pathways allow for triplet oxygen to add to the hydrocarbon directly, without the need for intersystem crossing. Thus, it appears that previously guessed mechanisms need to be revised. Further studies are under way in John Tully's group to incorporate these results into kinetic Monte Carlo simulations that will bridge the gap between atomic-scale reaction simulations and the macro-scale erosion process.

Conclusion

These examples indicate that an arsenal of simulation approaches is needed for nanoscale materials research including electrodynamics, empirical potential MD, direct dynamics, and CPMD. Fortunately, many of the necessary tools are available, or can be developed by straightforward modifications of existing software.

These examples also show that simulation is an important tool for many nanoscale materials problems. Although brute force simulation is not usually effective because the time scales are too long and the number of particles is too large, a combination of simulation in conjunction with analytical theory and simple models can be quite effective.

Acknowledgments

The research described here is supported by the AFOSR (MURI and DURINT programs) and the National Science Foundation (NSEC program). Important contributors to these projects have been my students and postdocs (Anne Lazarides, Guosheng Wu, Hai Long, Diego Troya, Ronald Pascual, Lance Kelly, and LinLin Zhao) as well as Tim Minton (Montana State) and Chad Mirkin (Northwestern).

DISTRIBUTED CYBERINFRASTRUCTURE SUPPORTING THE CHEMICAL SCIENCES AND ENGINEERING

Larry L. Smarr
University of California at San Diego

During the next 10 years, chemical science and engineering will be participating in a broad trend in the United States and across the world: we are moving toward a distributed cyberinfrastructure. The goal will be to provide a collaborative framework for individual investigators who want to work with each other or with industry on larger-scale projects that would be impossible for individual investigators working alone. I have been involved in examining the future of the Internet over the next decade, and in this paper I discuss this future in the context of the issues that were dealt with at the workshop.

Consider some examples of grand challenges that are inherently chemical in nature:

- simulating a micron-sized living cell that has organelles composed of millions of ribosomes, macromolecules with billions of proteins drawn from thousands of different species, and a meter of DNA with several billion bases—all involving vast numbers of chemical pathways that are tightly coupled with feedback loops by which the proteins turn genes on and off in regulatory networks; and

• simulating the star-forming regions in which radiation from newly forming stars causes a complex set of chemical reactions in diffuse media containing carbon grains as catalytic surfaces moving in turbulent flow. According to data from the Hubble Space Telescope, the size of the "reactor" would be on the order of one light year.

Both examples require solving a similar set of equations, although the regime in which they are applied is quite different from traditional chemical engineering applications.

Other areas of science have similar levels of complexity, and I would argue that those areas have progressed farther than the chemical sciences in thinking about the distributed information infrastructure that needs to be built to make the science possible.

• *Medical Imaging with MRI:* The NIH has funded the Biomedical Informatics Research Network (BIRN), a federated repository of three-dimensional brain images to support the quantitative science of statistical comparison of the brain subsystems. The task includes integration of computing, networks, data, and software as well as training people.

• *Geophysics and the Earth Sciences:* The NSF is beginning to roll out EarthScape, a shared facility which will ultimately have hundreds of high-resolution seismic devices located throughout the United States. The resulting data will provide three-dimensional images of the top several kilometers of the earth to the academic community.

• *Particle Physics:* The physics community is developing a global data Grid that will provide tens of thousands of users, in hundreds of universities, in dozens of countries throughout the world with a distributed collaborative environment for access to the huge amount of data generated at CERN in Geneva.

Many additional examples exist across scientific disciplines that have been involved as a community for 5 or 10 years in coming to consensus about the need for ambitious large-scale, data-intensive science systems. You will be hearing more and more about a common information infrastructure that underlies all of these—called distributed cyberinfrastructure—as it rolls out over the next few years. Here I want to impart a sense of what will be in that system. It seems to me that chemical engineering and chemistry share many aspects of data science with these other disciplines, so chemists and chemical engineers should be thinking about the problem.

For 20 years, the development of information infrastructure has been driven by Moore's Law—the doubling of computing speed per unit cost every 18 months.[1] But now, storage and network bandwidth are rising even faster than computing speed. This means that everything we thought we knew about integrated infrastructure will be turned completely upside down. It is why the Grid movement described earlier by Thom Dunning[2] is now taking off. We are going through a phase transition—from a very loosely coupled world to a very tightly coupled world—because the bandwidth between computers is growing much faster than the speed of the individual computers.

The rapid growth in data handling is being driven by the same kind of effect that influenced computing over the last 10 years: that is, not only did the individual processors become faster but the amount of parallelism within high-performance computers also increased. Each year or so, as each processor became twice as fast, you also used twice as many processors, thereby increasing the overall speed by a factor of 4—this leads to hyperexponential growth. We have entered a similar era for bandwidth that enables data to move through optical fibers. It is now possible to put wavelength bins inside the fiber, so that instead of just a single gigabit-per-second channel it is possible to have 16, 32, or many more channels (called lambdas), each operating at a slightly different wavelength. Every year both the number and the speed of these channels increase, thereby creating a bandwidth explosion. As an aside, the capability of optical fiber to support these multiple channels, ultimately enhancing the chemical sciences using the fiber, is itself a victory for chemical engineering. It was the ability, particularly through the contributions of Corning, to remove almost every molecule of water from the glass (water absorbs in the infrared wavelengths used by the fiber to transmit information) that makes this parallel revolution possible with optical fiber.

Without going into more detail, over the next few years each campus and state will be accumulating "dark fiber" to support such parallel dedicated networks for academic research or governmental functions. Each campus will have to start worrying about where the conduit is, so that you can pull the dark fiber from building to building. Once this owner-operated dark fiber is in place, you will get more bandwidth every year for no money other than for increasing the capability of the electronic boxes at the ends of the fiber. Consequently, you will use the increasing lambda parallelism to get an annual increase in bandwidth out of the same installed fiber. Today, this is occurring on a few campuses such as UCSD, as well as in some states, and there are discussions about building a national "light rail" of dark fiber across the country. When that happens, we will be able to link researchers together with data repositories on scales we haven't imagined.

Coming back to the science drivers, state-of-the-art chemical supercomputing

---

[1] Moore, G. E., *Electronics* **1965**, *38 (8)* 114-17; http://www.intel.com/research/silicon/mooreslaw.htm.
[2] See T. Dunning, Appendix D.

engineering simulations of phenomena such as turbulence (or experiments with laser read-outs of fluid) are comparable in scale to the highest resolution MRIs or earth seismic images, namely gigazone data objects. Such giant three-dimensional individual data objects ($10^3 \times 10^3 \times 10^3$, or larger) often occur in timed series, so one has a whole sequence of them. Furthermore the data are generated at a remote supercomputer or laboratory and stored there in federated data repositories. There may be some important consequences if you are the investigator:

- The data piece you want is somewhere you aren't, and you would like to interactively analyze and visualize these high-resolution and data objects.
- Because nature puts so many different kinds of disciplines together in what it does, you actually need a group of different kinds of scientists to do the analysis, and they are probably spread out in different locations, requiring collaborative networked facilities.
- Individual PCs don't have enough computing power or memory to handle these gigazone objects, requiring scalable Linux PC clusters instead.
- You won't have enough bandwidth over the shared Internet to interact with these objects; instead, you need dedicated lambdas.

You might ask, "What is the twenty-first century Grid infrastructure that is emerging?" I would answer that it is this tightly optically coupled Data Grid of PC clusters for computing and visualization with a distributed storage fabric, tied together a software middle layer that enables creation of virtual metacomputers and collaboration. In my institute we try to look out to see how this is all happening, by creating "Living Laboratories of the Future." We do this by deploying experimental networked test beds, containing bleeding edge technological components that will become mass market in 3, 5, or 10 years, and using the system now with faculty and students. That is the way supercomputers developed—from 15 years ago when a few people ran computational chemistry on a gigahertz machine like the Cray-2 that cost $20 million, to today, when everybody works on a gigahertz machine PC that costs only $1000.

Let me summarize a few major trends in this emerging distributed cyberinfrastructure:

- *Wireless Internet:* It took 30 years of exponential growth for the number of fixed Internet users, sitting at their PCs, to grow to two hundred million. We are likely to double that number with mobile Internet users who use cell phones or portable computers *in the next three years*. Consider that today there are probably a few hundred million automobiles in the United States, each having a few dozen microprocessors and several dozen sensors and actuators. When the processors on these cars get on the net they will vastly dominate the number of people in this country on the net. So, if you thought you had seen an explosion on the Internet, you really haven't seen anything yet.

- *Distributed Grid Computing:* Wisconsin's pioneering Condor[3] project, which creates a distributed computing fabric by tying hundreds of UNIX workstations together, was discussed during the workshop. Today, companies such as Entropia or United Devices are emulating that basic concept with PCs. In this sense, PCs are the dark matter of the Internet—there is an enormous level of computing power that is not being used but could be tied together by the Grid.
- *SensorsNets:* One of the problems with doing large-scale chemistry in the environment is that you don't know the state of the system. I think this may be generally true if you look at large-scale chemical plants. If you have wireless everywhere you can insert devices to measure the state of the system and obtain a finer-grained set of data. If you are simulating the future of the system, your simulation can be much more useful, and the approach will afford a tighter coupling between those who do simulation and those who gather data.
- *Miniaturization of Sensors:* One of the big trends that is taking place is the ability to take chemical sensing units—sensors that detect the chemical state of air, water, and so forth—and make them smaller and cheaper so you can put them in more places. I think chemistry and chemical engineering will be a frontier for SensorNets based on nanotechnology and wireless communication. Moreover, there is a revolution going on in the kinds of sensor platforms that can be used, such as automous vehicles the size of hummingbirds that fly around, gather, and transmit data.
- *Nanoscale Information Objects:* Consider the human rhinovirus, which has genetic information coded for the several surface proteins. These proteins form an interlocking coat that serves to protect the nucleic acid which is a software program—which, of course, is interpreted to generate the protein coat that protects the software. So, in addition to being a nano-object, the virus structure is also an information object as well. So when you talk about nanotechnology, I want you to remember that it is not just the physics, chemistry, and biology that is small, but there is a new set of nano information objects that are also involved.

For science, all of these trends mean that we are going to be able to know the state of the world in much greater detail than any of us ever thought possible—and probably on a much shorter time scale than most of us would expect as we focus on this report that looks at "the next 10 years."

---

[3]See: *http://www.cs.wisc.edu/condor/.*

MODELING AND SIMULATION AS A DESIGN TOOL[1,2]

Ellen Stechel
Ford Motor Company

The lens from which my comments arise is an industrial perspective, albeit from a limited experience base. The experience base reflected is one of four years in industrial research and development, during which about 60% of my time has been in research and 40% in development. Over those four years, my observations and experiences have caused my thinking to evolve dramatically.

The first section of this paper is focused on the chemical and materials simulation section in Ford Research Laboratory. I describe research that has significant academic character but simultaneously has a path of relevance to automotive applications. The second section focuses on some cross-cutting themes, and the third section describes three specific examples that illustrate those themes and reflect some challenges and opportunities for chemical sciences.

Chemical and Materials Simulation in an Automotive Industrial Research Laboratory

The unifying strategy of the chemical and materials simulation (CAMS) group at Ford Research Laboratory is the use of atomistic and higher length scale chemical and materials simulation methods to develop a knowledge base with the aim of improving materials, interfacial processes, and chemistry relevant to Ford Motor Company. Excellence and fundamental understanding in scientific research are highly valued; but relevance and potential impact are most important. There are three focus areas of research:

1. Surface chemistry and physics for such applications as environmental catalysis, exhaust gas sensors, mechanical wear and fatigue.
2. Materials physics, such as lightweight alloys and solid oxide fuel cells.
3. Molecular chemistry, which includes particulate matter dynamics, atmospheric chemistry, and combustion chemistry.

The tools used by the chemical and materials simulation group are commercial, academic, and homegrown simulation and analysis software, in addition to dedicated compute servers. The development of some of those "homegrown"

---

[1] Copyright reserved by Ford Motor Company. Reproduced by permission.
[2] Acknowledgement: The author is indebted to Drs. John Allison, Peter Beardmore, Alex Bogicevic, William Green, Ken Hass, William Schneider, and Chris Wolverton for many enlightening conversations and for use of their work to illustrate the points in this manuscript.

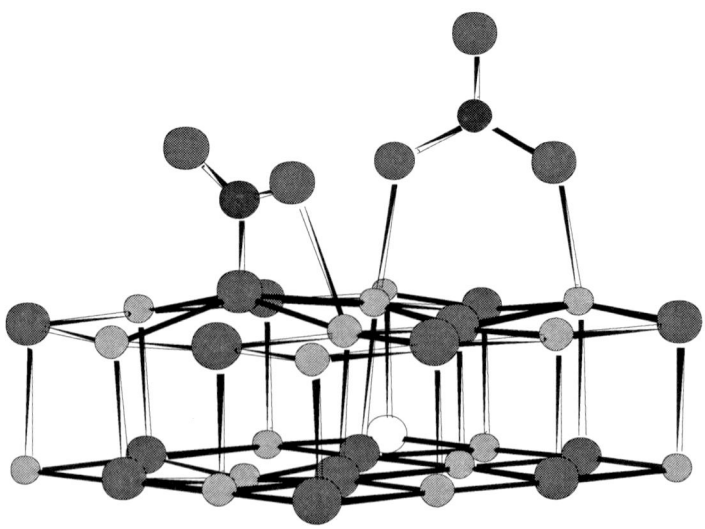

FIGURE 1 NO_x storage: Ab initio calculations of NO_x binding to an MgO surface provide a knowledge base.

tools occurred in non-industrial settings, since one of the proven ways to transfer technology is simply to hire a person, who often arrives with tools as well as expertise.

Catalytic Materials

Exhaust gas catalytic aftertreatment for pollution control depends heavily on chemistry and on materials. Hence, in one CAMS project, the goal is to guide the development of improved catalytic materials for lean-burn, exhaust-NO_x (NO and NO_2) aftertreatment, for gas sensors, and for fuel cells. One driver for improved materials is to enable cleaner lean-burn and diesel engine technologies. The first principles, atomistic-scale calculations have been able to elucidate NO_x binding to an MgO surface, providing a knowledge base relevant to the fundamentals of a NO_x storage device (Figure 1).[3] Using ab initio calculations, Ford researchers have been able to map out the energy landscape for NO_x decomposition and N_2O formation on a model Cu(111) surface, relevant to exhaust NO_x catalysis.[4]

---

[3] Schneider, W. F.; Hass, K. C.; Militec, M.; Gland, J. L. *J. Phys. Chem.* **2002,** *B 106,* 7405.
[4] Bogicevic, A.; Hass, K. C. *Surf. Sci.* **2002,** *506,* L237.

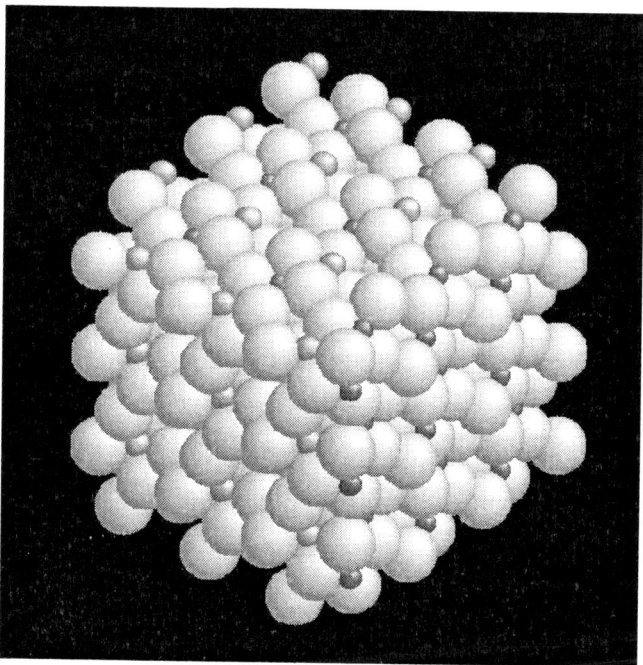

FIGURE 2 Ab initio calculations on yttria-stabilized zirconia to improve activity and robustness of oxide conductors for fuel cells and sensors.

Ionic Conductors

A second example from the CAMS group focuses on ionic conductors, with the goal of improving the activity and robustness of oxide conductors for fuel cells and sensors.[5] The scientific goal of the project is to understand with the intent of tailoring ion diffusion in disordered electrolytes. This work has developed a simple computational screen for predicting ionic conductivity of various dopants. The work has also established the only known ordered yttria-stabilized zirconia phase (Figure 2).

Cross-cutting Themes

While the two examples briefly described above have direct relevance to the automotive industry, the scientific work is indistinguishable from academic basic research. Nevertheless, successful industrial research generally reflects and builds

---

[5]Bogicevic A.; Wolverton, C. *Phys. Rev. B* **2003**, *67*, 024106; Bogicevic A.; Wolverton, C. *Europhys. Lett.* **2001**, *56*, 393; Bogicevic A.; Wolverton, C.; Crosbie, G.; Stechel, E. *Phys. Rev. B* **2001**, *64*, 014106.

on four recurring and broadly applicable themes. The first theme is the false dichotomy that arises when trying to distinguish applied from basic research. The second theme is a definition of multidisciplinary research that is broader than what is conventional. The application of this broader definition leads to the challenge of integration across research fields within the physical sciences as well as across science, engineering, technology, and business. The third theme concerns complex systems and requires establishing an appropriate balance between reductionism (solvable "bite-size" pieces) and holism (putting the pieces back together again). Finally, the fourth theme is one of a four-legged stool, a variation of the often-referenced three-legged stool. A successful application of industrial research necessarily adds a leg to the stool and puts something on the seat of the stool. The three legs are theory and modeling, laboratory experimentation, and computer simulation. The fourth leg is subsystem and full-system testing (verification and validation of laboratory concepts), and what sits on top of the stool is the real-world application in the uncontrolled customer's hands.

False Dichotomy

The structure of science and technology research sometimes is assumed to have a linear progression from basic research (which creates knowledge that is put in a reservoir) to application scientists and engineers (who try to extract knowledge from the reservoir and apply it for a specific need), and finally to develop technology. Few people believe that the science and technology enterprise actually works in this simplistic way. Nevertheless, with this mental model underpinning much of the thinking, unintentional adverse consequences result.

As an alternative mental model, Pasteur's Quadrant,[6] generalizes the linear model to a two-dimensional model, where one axis is consideration and use, and the other axis is the quest for fundamental understanding. The quadrant in which there is both a strong consideration of use and a strong quest for fundamental understanding ("Pasteur's Quadrant") has basic and applied research rolled together. This is the quadrant with the potential for the greatest impact in industrial research. The false dichotomy between basic and applied research arises when the one-dimensional model ignores feedback—where the application drives a need for new fundamental understanding that is not yet in the reservoir. Indeed, in implementing new technology, one is likely to find that what keeps one awake at night is not the known but the unknown—missing basic research that is applicable to the technology. Industrial researchers quickly find no shortage of challenging, intellectually stimulating, and poorly understood (albeit evolutionary) research problems that are relevant and have the potential for great impact on the industry.

---

[6]*Pasteur's Quadrant*, Stokes, D. E. Brookings Institution Press: Washington, DC, 1997.

Another way of looking at the two-way relationship between science and application is to recognize that the difference is in the starting and ending points, but not necessarily in the road traveled. If the primary motivation is science—a quest for understanding—then the world in which the researcher lives is one that is reasonably deterministic, predictive, and well controlled. The goal is generally one of proof of concept—a consistent explanation or reproducibility of observations. New and novel knowledge and wisdom, new capabilities, and new verification and validation methodologies are outputs of the science.

However, if the motivation stems from an industrial or application perspective, the researcher begins first by considering the use of what he or she is studying. As the researcher digs beneath the surface as if peeling away layers of an onion, he or she finds that it is necessary to ask new questions; this drives new science, which drives new knowledge, and, which in turn, drives new uses and new technology. Additionally, the two researchers (understanding-driven and application-driven) frequently meet in the middle of their journey—with both finding the need to develop new and novel knowledge and wisdom, new capabilities, and new verification and validation methodologies.

It does not matter where the researcher starts the journey: whether he or she comes from an industrial or academic perspective, there is common ground in the middle of the journey. In industry, science is a means to an end, but it is no less science. The world the industrial researcher strives for is a world defined by performance within specifications, robustness, reliability, and cost. However, in trying to achieve those goals, the researcher necessarily asks new questions, which drive new science in the process of deriving verifiable answers to those questions.

Integrated, Complex Systems, Perspective

To strike the appropriate balance between reductionist approaches and integrated approaches is a grand challenge. It may be that as social creatures, people are simply ill-equipped to go beyond reductionist approaches, to approaches that integrate science, engineering, technology, economics, business, markets, and policy. Right now, there is a strong tendency to view these disparate disciplines in isolation. Nevertheless, they interrelate—and interrelate strongly—in the real world.

Even when starting with a problem-driven approach, a goal or problem can always be broken down into solvable pieces. The very basis for the success of twentieth century science has been such reductionism. Nevertheless, to obtain optimal solutions, it is necessary to appropriately integrate disparate pieces together. Within any system, if one only optimizes subsystems, then one necessarily suboptimizes the entire system.

Typically, there will be many ways to assemble pieces. Industry needs a reassembled state that is practical, cost-effective, robust, adaptable, strategic, and so on. Although the constraints are many, it is necessary to consider all of them seriously and intellectually.

Complex Systems

The term *"complex systems"* means different things to different people. An operational definition is used here: complex systems have, nonlinear relationships, large experimental domains, and multiple interactions between many components. For complex systems, the underlying model is generally unknown, and there is no analytical formula for the response surface. It may be that solutions come in the form of patterns rather than numerical values. Most importantly, the whole is not the sum of the parts.

A system can be very complicated and can be ordered, or at least deterministic. One might be able to write down a model and use a very powerful computer to solve the equations. Science can make very good progress in elucidating solutions of such a system. A system can also be random enough that many variables may not be explicitly important, but their effects can be adequately accounted for with statistical science. It is important not to ignore such effects, but often they are describable as noise or random variables.

Somewhere between the ordered state and the fully random state lie many complex systems of interest. Significant progress depends on both the understanding-driven and the application-driven researcher comprehending and becoming comfortable with such complexity.

The Four-Legged Stool and Statistics

Theory, experiment, and simulation form a three-legged stool. The addition of subsystem and full-system testing produces a four-legged stool. The ultimate goal is to have a firm grasp of how a product will perform in the customer's hands, and that is what sits on top of the stool. The four legs must support the development of a firm understanding of mean performance variables, natural variability, and the sources of variability as well as the failure modes and the ability or not to avoid those failure modes. One does not need to be able to predict performance under all conditions (something which is often impossible), but one does need the ability to determine if the anticipated stresses can or cannot lead to a hard failure (something breaks) or a soft failure (performance degrades to an unacceptable level.)

Consequently, it is critically important that the four legs do not ignore statistical performance distributions. Statistical science is a powerful tool for comprehending complex systems; however, it does not appear to be a well-appreciated discipline, despite its importance to product development. In addition, of note are the tails of performance distributions, which become quite significant in a population of millions for high-volume products.

Examples

Three automotive examples that can and do benefit from simulation and from the chemical sciences are aluminum casting for engine blocks and heads, exhaust aftertreatment catalysts for air quality, and homogeneous charged compressive ignition (HCCI) engines—a potential future technology for improving fuel economy and maintaining low emissions. In the following discussion, the focus for each example is on why the industry has an interest in the application, what the industry is trying to accomplish, and the nature of some of the key ongoing scientific challenges. The emphasis is on how the chemical sciences fit in and how simulation fits in, but the challenge in each example involves significantly more than chemical sciences and simulation.

Virtual Aluminum Casting

Aluminum casting research at Ford Motor Company is the most integrated of my three examples. I chose it because this research illustrates good working relations across disciplines and research sectors. The research in the aluminum-casting project spans the range from the most fundamental atomic-scale science to the most applied in the manufacturing plant, and the project exploits all legs of the four-legged stool. The research approach displays a heavy reliance on simulation that unimpeachably is having significant impact.

Modeling and simulation of *new* materials receives most of the attention these days, but breakthroughs on mature material may have greater potential impact. Peter Beardmore, a now-retired director from Ford Research Laboratory, argued [Ref. P. Beardmore, NAE, 1997] that refinements made to mature materials and processes would have a much greater probability of being implemented—and therefore would have far greater impact—than all new materials concepts combined. While his focus was on automotive structural materials, I believe the statement is broadly applicable and is likely true of virtually all technology.

Automobiles are mostly steel, by weight. However, the automotive industry comprises a large percentage of the consumer market for aluminum, iron, and magnesium. The automotive industry does not dominate the plastics market.

Materials usage (measured by weight) did not changed much from 1977 to 2001. Vehicles are still mostly steel, which accounts for the high recyclability of the automobile. Aluminum and plastic content have increased some. However, with emerging energy issues, further change will be necessary. Fuel economy and emission standards are very strong drivers for lightweight materials, since to a first approximation fuel economy is inversely proportional to the weight of a vehicle. Any other avenues to increase fuel economy pale in terms of what arises from lightening the vehicle. However, from a consumer perspective, it is critical that the vehicle is reliable, cost-effective, and high quality: consumer demands thus present considerable challenges to lightweight materials.

To attack the problem from a technology perspective, the researcher must consider the material's usage, performance, reliability, and manufacturability. The automotive industry makes a large number of cylinder heads and engine blocks; so reducing weight creates a driver to replace cast iron with cast aluminum. To have impact, the researcher must set goals and meet specifications, such as reducing product and process development time by a large percentage through simulation, improving quality, reducing scrap, and ensuring high-mileage durability.

The Ford Research virtual aluminum-casting project has developed unique capabilities for simultaneous process and product optimization. Generally, process optimization and product optimization are not integrated. In the stereotypical process, a design engineer generates a product design, and then throws it over the fence to the manufacturing division. The manufacturing engineer says, "Well, if you made this change in your design it would be easier to manufacture," and an argument begins. Both parties remain focused on their subsystem and neither is able to recognize that the design and manufacturability of the product is a system, and that system optimization as opposed to subsystem optimizations is often where opportunities emerge.

The virtual aluminum-casting project has developed a comprehensive and integrated set of simulation tools that capture expertise in aluminum metallurgy casting and heat treatment—years of experience captured in models. In addition to enabling calculations, an often-overlooked characteristic of models, is that they can organize knowledge and information and thereby assist in knowledge dissemination. The models in the toolset have unsurpassed capabilities. John Allison, the project leader for virtual aluminum casting uses the phrase "computational materials science *plus*." The project, which links physics, materials science, mechanics, theory, modeling, experiment, engineering, fundamental science, and applied science, is achieving notable successes.

The structure of this virtual aluminum casting project could serve as a readily generalized example. One of the fundamental premises is that computational models should be able to solve 80 percent of the problem quickly through simulation, and the models should not be trusted completely. Experimental verification and validation on subsystem and full-system levels are critical. Another fundamental premise is that new knowledge should be transferred into the plant and implemented. In other words, it is important to go beyond proof of concept to practical application.

The virtual aluminum casting project is comprehensive and multiscale;[7] it reaches from initial product design based on property requirements, all the way to influencing actual manufactured components (Figure 3). The project involves experiments and models at all length scales and all levels of empiricism that predict properties, service life, damage, stress, and the like.

---

[7]Vaithyanathan, V.; Wolverton, C.; Chen, L.-Q. *Phys. Rev. Lett.* **2002**, *88*, 125503.).

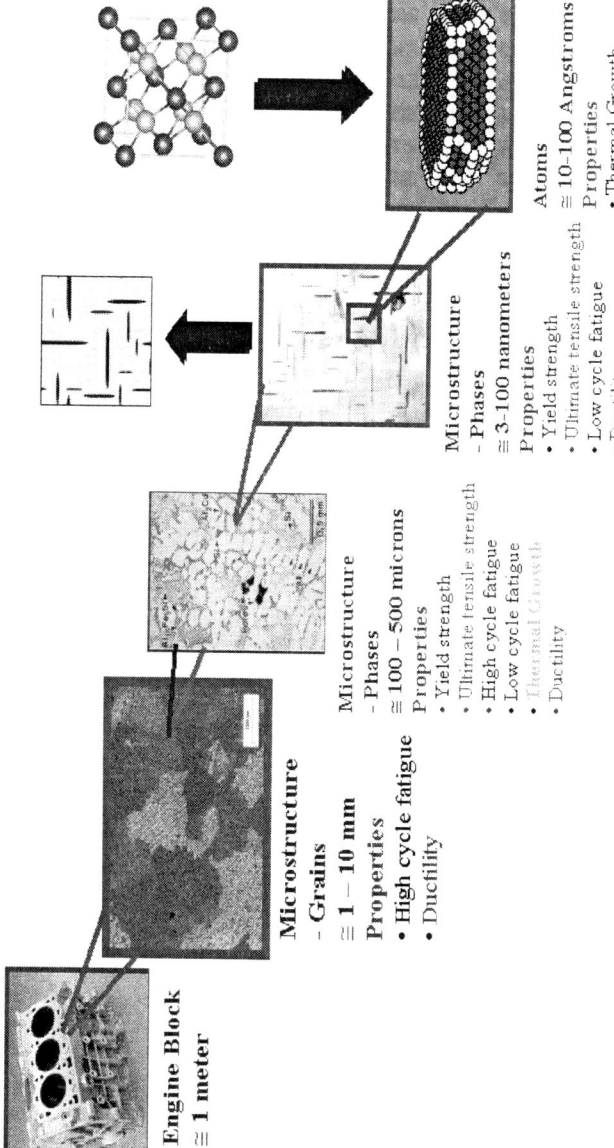

FIGURE 3 Multiscale calculations of aluminum alloys to guide the development of improved engine blocks.

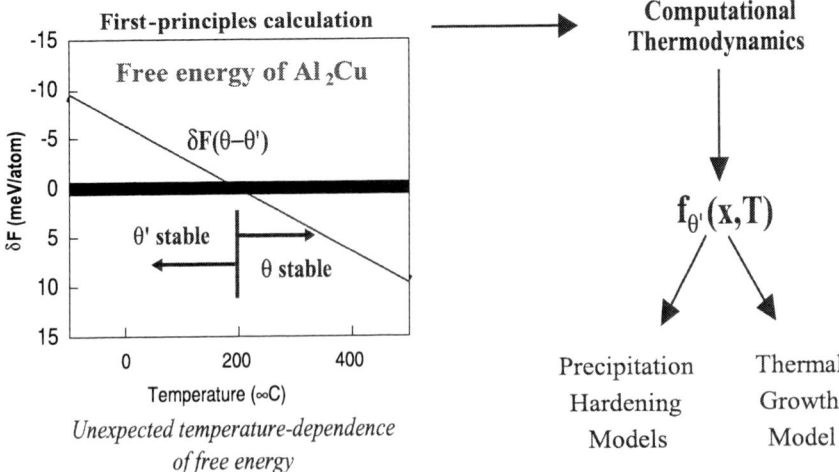

FIGURE 4   First-principles calculations can provide input to models downstream.

The aluminum alloy material has roughly 11 components (as a recycled material, the alloy actually has an unknown number of components that depend on the recycling stream), and some of the concentrations are quite small. Some components have more impact than others do. Impurities can have enormous effects on the microstructure and thereby affect the properties. Precipitates form during heat treatment, and manufacturers have the opportunity to optimize the aluminum heat treatment to achieve an optimized structure. First principles calculations and multiscale calculations are able to elucidate opportunities, including overturning 100 years of metallurgical conventional wisdom (Figure 4).[8]

Several key challenges must be overcome in this area. In the chemical sciences, methodologies are needed to acquire quantitative kinetic information on real industrial materials without resorting to experiment. It is also necessary to determine kinetic pre-factors and barriers that have enough accuracy to be useful. Another key challenge is the need for seamless multiscale modeling including uncertainty quantification. Properties such as ductility and tensile strength are still very difficult to calculate. Finally, working effectively across the disciplines still remains a considerable barrier.

---

[8]Wolverton, C.; Ozolins, V. *Phys. Rev. Lett.* **2001,** *86,* 5518; Wolverton, C.; Yan, X.-Y.; Vijayaraghavan, R.; Ozolins, V. *Acta Mater.* **2002,** *50,* 2187.)

Catalysis of Exhaust Gas Aftertreatment

Exhaust aftertreatment catalysts provide a second, less-integrated example of the use of chemical sciences modeling and simulation to approach an application-related problem. Despite large investments in catalysis, both academia and funding agencies seem to have little interest in catalysis for stoichiometric exhaust aftertreatment. Perhaps the perception of a majority of the scientific community is that stoichiometric exhaust aftertreatment is a solved problem. Although there is a large empirical knowledge base and a cursory understanding of the principles, from my vantage point the depth of understanding is far from adequate especially considering the stringency of Tier 2 tailpipe emissions regulations. This is another challenging and intellectually stimulating research area with critical issues that span the range from the most fundamental to the most applied.

After the discovery in the 1960s that tailpipe emissions from vehicles were adversely affecting air quality, there were some significant technological breakthroughs, and simply moving along a normal evolutionary path, we now have obtained three-way catalysts that reduce NO_x and oxidize hydrocarbons and CO at high levels of efficiency with a single, integrated, supported-catalyst system. Unfortunately, the three-way catalyst works in a rather narrow range of air-fuel ratio, approximately the stoichiometric ratio. This precludes some of the opportunities to improve fuel economy—for example, a leaner air-fuel ratio can yield better fuel economy but, with current technology, only at the expense of greater NO_x emissions. Also, most exhaust pollution comes out of the tailpipe in the first 50 seconds of vehicle operation, because the catalyst has not yet reached a temperature range in which it is fully active.

The three-way catalyst is composed of precious metals on supported alumina with ceria-based oxygen storage, coated on a high-surface-area ceramic or a metallic substrate with open channels for the exhaust to pass through and over the catalyst. As regulations become increasingly stringent for tailpipe emissions, the industry is transitioning to higher cell densities (smaller channels) and thinner walls between channels. This simultaneously increases the surface area, decreases the thermal mass, and reduces the hydraulic diameter of the channels; all three effects are key enablers for higher-efficiency catalysts. The high-surface-area nano-structured washcoat is another key enabler, but it is necessary to maintain the high surface area of the coating and highly dispersed catalytically active precious metals for the life of the vehicle despite high-temperature operation and exposure to exhaust gas and other chemical contaminants. In other words, the material must have adequate thermal durability and resist sintering and chemical poisoning from exhaust gas components. Neither thermal degradation nor chemical degradation is particularly well understood beyond some general principles; i.e., modeling is difficult without a lot of empirical input.

The industry also must design for end of life, which is a very large design space. Simulation currently uses empirically derived, aged-catalyst-state input.

What the industry really needs is a predictive capability. The industry also has to minimize the cost and the amount of the active catalytic platinum, palladium, and rhodium used, since precious metals are a rare and limited commodity.

Again a key chemical science challenge is kinetics. How does one obtain quantitative kinetic information for real industrial, heterogeneous catalysts without resorting to time-consuming experiments? Simulations currently use empirically derived simplified kinetics.

The science of accelerated aging has generally been ignored. The automotive industry must be able to predict aging of catalysts and sensors without running many vehicles to 150,000 miles. The industry does currently take several vehicles and accumulate miles, but driving vehicles to that many miles is a particularly slow way to get answers and does not provide good statistics. The industry does utilize accelerated aging, but it is done somewhat empirically without the benefit of a solid foundation of basic science.

Homogeneous Charge Compression Ignition

The third and final example arises from the Ford Motor Company–MIT Alliance, a partnership that funds mutually beneficial collaborative research. The principal investigators, Bill Green and Bill Kaiser, are from MIT and Ford, respectively. The project spans a range from the most fundamental chemistry and chemical engineering to the most applied. In contrast to the first two examples, this project focuses on a technology in development, as opposed to a technology already in practice.

Homogeneous charged compression ignition is similar in concept to a diesel engine. It is high efficiency and runs lean. It is compression ignited, as opposed to spark ignited, but it has no soot or NO_x because it runs much cooler than diesel. It is similar to a gasoline engine in that it uses pre-mixed, volatile fuels like gasoline, and it has similar hydrocarbon emissions. But an HCCI engine has a much higher efficiency and much lower NO_x emissions than a gasoline engine, which could eliminate the need for the costly three-way precious metal catalyst.

However, HCCI is difficult to control. There is no simple timing mechanism, which can control ignition, as exists for a spark or diesel fuel injection engine. HCCI operates by chemistry and consequently is much more sensitive to the fuel chemistry than either spark-ignition or diesel engines. The fuel chemistry in an internal combustion engine largely ignores the details of the fuel. HCCI looks very promising, but researchers do not yet know what the full operating range is or how to reliably control the combustion with computation. Yelvington and Green have already demonstrated that HCCI can work well over a broad range of conditions demonstrating the promise that computation and simulation can and must play a continuing and large role in resolving the HCCI challenges.[9]

---

[9]Yelvington, P. E.; Green, W. H.; *SAE Technical Paper* **2003**, 2003-01-1092.

Final Words

The complexity of industrial design problems requires that one must be able to deal with the realistic, not overly idealized, system. The complexity demands a multidisciplinary approach and working in teams. Access to and understanding of the full range of methods are generally necessary if the researcher is going to have impact by solving relevant problems, and individual industrial researchers must be able to interact and communicate effectively beyond their disciplines. To achieve integration from fundamental science to real-world applications is truly a challenge of coordination, integration, and communication across disciplines, approaches, organizations, and research and development sectors.

In general, approximate answers or solutions with a realistic quantification of uncertainty—if arrived at quickly—have greater impact than highly accurate answers or solutions arrived at too late to impact critical decisions. Often, there is no need for the increased level of accuracy. To quote Einstein, "Things should be made as simple as possible, but not any simpler." Sometimes researchers in industry, just out of necessity, oversimplify. That is when the automotive development engineer might lose sleep, because it could mean that vehicles might have unexpected reliability issues in the field, if the oversimplification resulted in a wrong decision.

Simulations with varying degrees of empiricism and predictive capability should be aligned closely with extensive experimental capabilities. It is also important to bring simulation in at the beginning of a new project. Too often, experimentalists turn to a computational scientist only after repeated experimental failures. Many of these failures would have been avoided had the consultation occurred at an early stage. The computational expert could help generate hypotheses even without doing calculations or by doing some simple calculations, and the perspective of the computational researcher can frequently eliminate dead ends. In addition, framing questions correctly so that the experiments yield unambiguous results, or reasonably unambiguous results, is crucial. Obtaining reasonably definitive answers in a timely manner is equally crucial, but too often, there does not seem to be a time scale driving needed urgency.

Data and knowledge management offer additional overlooked opportunities. Hypotheses that have been proven wrong often continue to influence decisions. Decision makers too often operate on the basis of disproved or speculative conjectures rather than on definitive data. The science and technology enterprise needs a way to manage data such that it is relatively easy to know what the community does and does not know as well as what the assumptions are that underlie current knowledge.

Finally, it is important to have realistic expectations for success. This is a difficult challenge because what one measures and rewards is what one gets. There are many ways to impact applications and technology with any level of sophistication in a simulation. Some of the important ways lend themselves only to intangible measures, but oftentimes these may be the most important. Again quoting Einstein, "Not everything that counts can be counted, and not everything that can be counted counts."

DRUG DISCOVERY: A GAME OF TWENTY QUESTIONS

Dennis J. Underwood
Infinity Pharmaceuticals, Inc.

Introduction

There is a revolution taking place in the pharmaceutical industry. An era in which nearly continuous growth and profitability is taken for granted is coming to an end. Many of the major pharmaceutical companies have been in the news lately with a plethora of concerns ranging from the future of the biotechnology sector as a whole to concerns over the availability of drugs to economically disenfranchised groups in the developed and the developing world. The issues for the pharmaceutical sector are enormous and will likely result in changes in the health-care system, including the drug discovery and development enterprises. Central challenges include the impact that drugs coming off patents have had on the recent financial security of the pharmaceutical sector and the need for improved protection of intellectual property rights. Historically, generic competition has slowly eroded a company's market share. There is a constant battle between the pace of discovering new drugs and having old drugs going off patent, providing generic competition opportunities to invade their market share. Recent patent expirations have displayed a much sharper decline in market share, making new drugs even more critical.

The 1990s were a decade in which the pharmaceutical giants believed they could sustain growth indefinitely by dramatically increasing the rate of bringing new medicines to market simply by increasing R&D spending and continuing to utilize the same research philosophies that worked in the past. It is clear from the rapid rise in R&D expenditure and the resultant cost of discovering new drugs that the "old equation" is becoming less favorable. There is a clear need to become more efficient in the face of withering pipelines and larger and more complex clinical trials. For large pharmaceutical companies to survive, they must maintain an income stream capable of supporting their current infrastructure as well as funding R&D for the future. The cost of drug development and the low probability of technical success call for improved efficiency of drug discovery and development and further investment in innovative technologies and processes that improve the chances of bringing a compound to market as a drug. Already there has been quite a change in the way in which drugs are discovered. Large pharmaceutical companies are diversifying their drug discovery and development processes: They are relying more on the inventiveness of smaller biotechnology companies and licensing technology, compounds, and biologicals at a faster, more competitive rate. To meet critical time lines they are outsourcing components of research and development to contract research organizations, enabling greater efficiencies by providing added expertise or resources and decreasing development time lines. The trend toward mergers and acquisitions, consolidating pipelines, and attempting to achieve economies of scale is an attempt by large pharmaceutical companies to build competitive organizations. Although this may help short-term security, future ongoing success may not be ensured solely with this strategy.

One of the most valuable assets of a pharmaceutical company is its experience in drug discovery and development. Of particular importance are the data, information and knowledge generated in medicinal chemistry, pharmacology, and in vivo studies accumulated over years of research in many therapeutic areas. This knowledge is based on hundreds of person-years of research and development; yet most companies are unable to effectively capture, store, and search this experience. This intellectual property is enormously valuable. As with the other aspects of drug discovery and development, the methods and approaches used in data-, information- and knowledge-base generation and searching are undergoing evolutionary improvements and, at times, revolutionary changes. It is imperative for all data- and information-driven organizations to take full advantage of the information they are generating. We assert that those companies that are able to do this effectively will be able to gain and sustain an advantage in a highly complex, highly technical, and highly competitive domain. The aim of this overview is to highlight the important role informatics plays in pharmaceutical research, the approaches that are currently being pursued and their limitations, and the challenges that remain in reaping the benefit of advances. We are using the term "informatics" in a general way to describe the processes whereby information is generated from data and knowledge is derived as our understanding builds. Informatics also refers to the management and transformation of data, information, and assimilation of knowledge into the processes of discovery and development.

There has been much time, money, and effort spent in attempting to reduce the time it takes to find and optimize new chemical entities. It has proven difficult to reduce the time it takes to develop a drug, but the application of new technologies holds hope for dramatic improvements. The promise of informatics is to reduce development times by becoming more efficient in managing the large amounts of data generated during a long drug discovery program. Further, with managed access to all of the data, information, and experience, discoveries are more likely and the expectation is that the probability of technical success will increase.

Why is the probability of technical success of drug discovery and development so low? What are the issues in moving compounds through the drug discovery and development pipeline? It is clear from existing data that the primary bottlenecks are pharmacokinetic problems and lack of efficacy in man. In addition there are problems of toxicity in animal models and the discovery of adverse effects in man. Clearly there are significant opportunities to improve the probability of technical success and, perhaps, to shorten the time line for development. The current strategies bring in absorption, distribution metabolism, excretion and toxicity (ADME-Tox) studies earlier in the process (late discovery) and use programs based on disease clusters rather than a single target.[1]

Drug discovery and development is a difficult business because biology and the interaction with biology is complicated and, indeed, may be classifiable as a complex system. Complexity is due partly to an incomplete knowledge of the biological components of pathways; the manner in which the components interact and are compartmentalized; and the way they are modulated, controlled, and regulated in response to intrinsic and environmental factors over time. Mostly, biological interactions are important and not accounted for in screening and assaying strategies. Often model systems are lacking in some way and do not properly represent the target organism. The response of the cell to drugs is very dependent on initial conditions, which is to say that the response of the cell is very dependent on its state and the condition of many subcellular components. Typically, the behavior of complex systems is different from those of the components, which is to say that the processes that occur simultaneously at different scales (protein, nucleic acid, macromolecular assembly, membrane, organelle, tissue, organism) are important and the intricate behavior of the entire system is dependent on the processes but in a nontrivial way.[2,3] If this is true, application of the tools of reductionism may not provide us with an understanding of the responses of an organism to a drug. Are we approaching a "causality catastrophe" whenever we expect the results of in vitro assays to translate to clinical data?

What is an appropriate way to deal with such complexity? The traditional approach has been to generate large amounts of replicate data, to use statistics to provide confidence, and to move cautiously, stepwise, toward higher complexity: from in vitro to cell-based to tissue-based to in vivo in model animals and then to man. In a practical sense, the drug discovery and development odds have been improved by a number of simple strategies: Start with chemically diverse leads, optimize them in parallel in both discovery and later development, back them up with other compounds when they reach the clinic and follow on with new, structurally different compounds after release. Approach diseases by focusing, in parallel, on clusters rather than on a single therapeutic target, unless the target has proven to be clinically effective. Generate more and better-quality data focusing on replicates, different conditions, different cell types in different states, and dif-

---

[1]Kennedy, T. *Drug Discovery Today* **1997**, *2 (10)*, 436-444.
[2]Vicsek, T. *Nature* **2001**, *411*, 421.
[3]Glass, L. *Nature* **2001**, *410*, 277-84.

ferent model organisms. Develop good model systems, such as for ADME-Tox, and use them earlier in the discovery process to help guide the optimization process away from difficulties earlier (fail early). The increase in the amount and diversity of the data leads to a "data catastrophe" in which our ability to fully extract relationships and information from the data is diminished. The challenge is to provide informatics methods to manage and transform data and information and to assimilate knowledge into the processes of discovery and development; the issue is to be able to integrate the data and information from a variety of sources into consistent hypotheses rich in information.

The more information, the better. For example, structure-based design has had many successes and has guided the optimization of leads through a detailed understanding of the target and the way in which compounds interact. This has become a commonly accepted approach, and many companies have large, active structural biology groups participating in drug discovery teams. One of the reasons that this is an accepted approach is that there is richness in data and information and there is a wealth of methodology available, both experimental and computational, that enables these approaches. The limitations in this area are concerned primarily with the challenges facing structural biology such as appropriate expression systems, obtaining sufficient amounts of pure protein, and the ability to crystallize the protein. These limitations can be severe and can prevent a structural approach for many targets, especially membrane-bound protein complexes. There is also a question of relevancy: Does a static, highly ordered crystalline form sufficiently represent dynamic biological events?

There are approaches that can be used in the absence of detailed structural information. Further, it is often instructive to use these approaches in conjunction with structural approaches, with the aim of providing a coherent picture of the biological events using very different approaches. However, application of these methods is extremely challenging primarily due to the lack of detailed data and information on the system under study. Pharmacophore mapping is one such approach. A complexity that is always present is that there are multiple ways in which compounds can interact with a target; there are always slight changes in orientation due to differences in chemical functionality and there are always distinctly different binding modes that are possible. Current methods of pharmacophore mapping find it difficult to detect these alternates. Further, the approach often relies on high-throughput screening approaches that are inherently "noisy" making the establishment of consistent structure activity relationships difficult. In addition, the methodology developed in this area is often limited to only parts of the entire dataset. There has been much effort directed to deriving two-, three-, and higher-dimensional pharmacophore models, and the use of these models in lead discovery and development is well known.[4]

---

[4]Agrafiotis, D. K.; Lobanov, V. S.; Salemme, F. R. *Nat. Rev. Drug Discov.* **2002**, *1*, 337-46.

The key issue in these methods is the manner in which compounds, their structure, features, character and conformational flexibility are represented. There are many ways in which this can be achieved, but in all cases the abstraction of the chemistry for computational convenience is an approximation. Each compound, in a practical sense, is a question that is being asked of a complex biological system. The answer to a single question provides very little information; however, in an manner analogous to the game of "twenty question," the combined result from a well-chosen collection of compounds (questions) can provide an understanding of the biological system under study.[5] The manner in which chemistry is represented is essential to the success of such a process. It is akin to posing a well-informed question that, together with other well-formed questions, is powerful in answering or giving guidance to the issues arising from drug discovery efforts. Our approach is to use the principles of molecular recognition in simplifying representation of the chemistry: atoms are binned into simple types such as cation, anion, aromatic, hydrogen-bond acceptor and donor, and so forth. In addition, the conformational flexibility of each molecule is represented. The result is a matrix in which the rows are compounds and the columns are bits of a very large string (tens of millions of bit long) that mark the presence of features. Each block of features can be the simple presence of a functional group such as phenyl, chlorophenyl, piperidyl, or aminoethyl, or it can be as complex as three- or four-pint pharmacophoric features that combine atom types and the distances between them. The richness of this representation along with a measure of the biological response of these compounds enables methods that use Shannon's maximum-entropy, information-based approach to discover ensembles of patterns of features that describe activity and inactivity. These methods are novel and have been shown to capture the essence of the effects of the SAR in a manner that can be used in the design of information-based libraries and the virtual screening of databases of known or realizable compounds.[6]

The power of these methods lies in their ability to discern relationships in data that are inherently "noisy." The data are treated as categorical rather that continuous: active (yes), inactive (no) and a variable category of unassigned data (maybe). These methods are deterministic and, as such, derive all relationships between all compounds at all levels of support. The relationships or patterns are scored based on their information content. Unlike methods such as recursive partitioning, pattern discovery is not "greedy" and is complete. The discrimination ability of pattern discovery depends very much on the quality of the data generated and on the type and condition of the compounds; if either is questionable, the

---

[5]Underwood, D .J. Biophys. J. **1995**, 69, 2183-4.
[6]Beroza, P.; Bradley, E. K.; Eksterowicz, J. E.; Feinstein, R., Greene, J.; Grootenhuis, P. D.; Henne, R. M.; Mount, J.; Shirley, W. A.; Smellie, A.; Stanton, R. V.; Spellmeyer, D. C. *J. Mol. Graph. Model.* **2002**, *18*, 335-42.

signal-to-noise ratio will be reduced and the quality of the results will be jeopardized. These approaches have been used in the study of G-protein Coupled Receptors[7,8] and in ADME-Tox studies.[9,10]

These methods have also been used in the identification of compounds that are able to present the right shape and character to a defined active site of a protein[11,12] In cases where the protein structure is known and the potential binding sites are recognized, the binding site can be translated into a bit-string that is in the same representational space as described above for the compounds. This translation is done using methods that predict the character of the space available for binding compounds. The ability to represent both the binding site(s) and compounds in the same way provides the mechanism to discriminate between compounds that are likely to bind to the protein. These approaches have been used for serine proteases, kinases and phosphatases.[13]

The game of 20 questions is simple but, almost paradoxically, has the ability to give the inquisitor the ability to read the subject's mind. The way in which this occurs is well understood; in the absence of any information there are many possibilities, a very large and complex but finite space. The solution relies on a tenet of a dialectic philosophy in which each question provides a thesis and an antithesis that is resolved by a simple yes or no answer. In so doing, the number of possibilities are dramatically reduced and, after 20 questions, the inquisitor is usually able to guess the solution. The solution to a problem in drug discovery and development is far more complex than a game of 20 questions and should not be trivialized. Even so, the power of discrimination through categorization of answers and integration of answers from diverse experiments provides an extremely powerful mechanism for optimizing to a satisfying outcome—a potential drug.

These approaches have been encoded into a family of algorithms known as pattern discovery (PD).[14] PD describes a family of novel methods in the category of data mining. One of the distinguishing features of PD is that it discovers rela-

---

[7]Wilson, D. M.; Termin, A. P.; Mao, L.; Ramirez-Weinhouse, M. M.; Molteni, V.; Grootenhuis, P. D.; Miller, K.; Keim, S.; Wise, G. *J. Med. Chem.* **2002**, *45*, 2123-6.

[8]Bradley, E. K.; Beroza, P.; Penzotti, J. E.; Grootenhuis, P. D.; Spellmeyer, D. C.; Miller, J. L. *J. Med. Chem.* **2000**, *43*, 2770-4.

[9]Penzotti, J. E.; Lamb, M. L.; Evensen, E.; Grootenhuis, P. D. *J. Med. Chem.* **2002**, *45*, 1737-40.

[10]Clark, D. E.; Grootenhuis, P. D. *Curr. Opin. Drug Discov. Devel.* **2002**, *5*, 382-90.

[11]Srinivasan, J.; Castellino, A.; Bradley, E. K.; Eksterowicz, J. E.; Grootenhuis, P. D.; Putta, S.; Stanton, R. V. *J. Med. Chem.* **2002**, *45*, 2494-500.

[12]Eksterowicz, J. E.; Evensen, E.; Lemmen, C.; Brady, G. P.; Lanctot, J. K.; Bradley, E. K.; Saiah, E.; Robinson, L. A.; Grootenhuis, P. D.; Blaney, J. M. *J. Mol. Graph. Model.* **2002**, *20*, 469-77.

[13]Rogers, W. T.; Underwood, D. J.; Argentar, D. R.; Bloch K. M.; Vaidyanathan, A. G. *Proc. Natl. Acad. Sci. U.S.A.*, submitted.

[14]Argentar, D. R.; Bloch, K. M.; Holyst, H. A.; Moser, A. R.; Rogers, W. T.; Underwood, D. J.; Vaidyanathan, A. G.; van Stekelenborg, J. *Proc. Natl. Acad. Sci. U.S.A.*, submitted.

tionships between data rather than relying on human interpretation to generate a model as a starting point; this is a significant advantage. Another important advantage of PD is that it builds models based on ensembles of inputs to explain the data and therefore has an advantage in the analysis of complex systems (such as biology[2,3]). We have developed a novel approach to PD that has been applied to bio-sequence, chemistry, and genomic data. Importantly, these methods can be used to integrate different data types such as those found in chemistry and biology. PD methods are quite general and can be applied to many different areas such as proteomics, text, etc.

Validation of these methods in bio-sequence space has been completed using well-defined and well-understood systems such as serine proteases[13] and kinases. PD in bio-sequence space provides a method for finding ensembles of patterns of residues that form a powerful description of the sequences studied. The similarity between patterns expresses the evolutionary family relationships. The differences between patterns define their divergence. The patterns express key functional and structural motifs that very much define the familial and biochemical character of the proteins. Embedded in the patterns are also key residues that have particular importance with respect to the function or the structure of the protein. Mapping these patterns onto the x-ray structures of serine proteases and kinases indicates that the patterns indeed are structurally and functionally important, and further, that they define the substrate-binding domain of the proteins. This leads to the compelling conclusion that since the patterns describe evolutionary changes (divergence and convergence) and also describe the critical features of substrate binding, the substrate is the driver of evolutionary change.[13]

A particular application of PD is in the analysis of variations of genetic information (single nucleotide polymorphisms, or SNPs). Analysis of SNPs can lead to the identification of genetic causes of diseases, or inherited traits that determine differences in the way humans respond to drugs (either adversely or positively). Until now, the main method of SNP analysis has been linkage disequilibrium (LD), which seeks to determine correlations among specific SNPs. A key limitation of LD however is that only a restricted set of SNPs can be compared. Typically SNPs within a local region of a chromosome or SNPs within genes that are thought to act together are compared. PD on the other hand, through its unique computational approach, is capable of detecting *all* patterns of SNPs, regardless of the genomic distances between them. Among these will be patterns of SNPs that are responsible for the disease (or trait) of interest, even though the individual SNPs comprising the pattern may have no detectable disease (or trait) correlation. This capability will greatly accelerate the exploitation of the genome for commercial purposes.

E

Biographies of Workshop Speakers

Charles H. Bennett is an IBM fellow at IBM Research, where he has worked on various aspects of the relation between physics and information. He received his bachelor's degree from Brandeis University, majoring in chemistry, and his Ph.D. from Harvard in 1970 for molecular dynamics studies (computer simulation of molecular motion). His research has included work on quantum cryptography, algorithmic information theory, and "quantum teleportation." He is an IBM fellow, a fellow of the American Physical Society, and a member of the National Academy of Sciences.

Anne M. Chaka is the group leader for computational chemistry at the National Institute of Standards and Technology in Gaithersburg, Maryland. She received her B.A. in chemistry from Oberlin College, her M.S. in Clinical Chemistry from Cleveland State University, and her Ph.D. in theoretical chemistry from Case Western Reserve University. In 1999-2000, she was Max-Planck-Society Fellow at the Fritz-Haber-Institut in Berlin. She spent 10 years at the Lubrizol Corporation as head of the computational chemistry and physics program and previously was technical director of ICN Biomedicals, Inc., an analytical research chemist for Ferro Corporation, and a Cray programming consultant to Case Western Reserve University for the Ohio Supercomputer Center. Active areas of her research include atomistic descriptions of corrosion, pericyclic reaction mechanisms, free-radical chemistry, heterogeneous and homogeneous catalysis, thermochemistry, and combustion and oxidation.

Juan J. De Pablo is H. Curler Distinguished Professor of Chemical Engineering at the University of Wisconsin-Madison. He received his B.S. from Universidad Nacional Autónoma de México and his Ph.D. from the University of California at Berkeley. His research interests include thermodynamics, phase equilibria, statistical mechanics, molecular modeling and simulation, and polymer physics. Molecular simulations play an important role in his research, in which he uses powerful methods and advanced computational techniques to study molecular motion and to probe the microstructure of fluids and solids with the aim of explaining and predicting macroscopic behavior. Professor de Pablo has published 50 journal articles in the areas of polymer physics, molecular simulations, thermodynamics, and statistical mechanics. He has received the National Science Foundation's National Young Investigator Award.

Thom H. Dunning, Jr., is Distinguished Professor of Chemistry and Chemical Engineering at the University of Tennessee (UT) and Distinguished Scientist in Computer Science and Mathematics at Oak Ridge National Laboratory (ORNL). He is also director of the Joint Institute for Computational Sciences, which was established by UT and ORNL to create advanced modeling and simulation methods and computational algorithms and software to solve the most challenging problems in science and engineering. He has authored nearly 150 scientific publications on topics ranging from advanced techniques for molecular calculations to computational studies of the spectroscopy of high-power lasers and the chemical reactions involved in combustion. Dr. Dunning received his B.S. in chemistry in 1965 from the University of Missouri-Rolla and his Ph.D. in chemical physics from the California Institute of Technology in 1970. He was awarded a Woodrow Wilson Fellowship in 1965-1966 and a National Science Foundation Fellowship in 1966-1969. He is a fellow of the American Physical Society and of the American Association for the Advancement of Science.

Christodoulos A. Floudas is professor of chemical engineering at Princeton University, associated faculty in the Program of Applied and Computational Mathematics at Princeton University, and associated faculty in the Department of Operations Research and Financial Engineering at Princeton University. He earned his B.S.E. at Aristotle University of Thessaloniki, Greece, and his Ph.D. at Carnegie Mellon University. He has held visiting professor positions at Imperial College, England; the Swiss Federal Institute of Technology; the University of Vienna, Austria; and the Chemical Process Engineering Research Institute (CPERI), Thessaloniki, Greece. His research interests are in the area of chemical process systems engineering and lie at the interface of chemical engineering, applied mathematics, operations research, computer science, and molecular biology. The principal emphasis is on addressing fundamental problems in process synthesis and design, interaction of design and control, process operations, discrete-continuous nonlinear optimization, deterministic global optimization, and computational chemistry, structural biology and bioinformatics. He is the recipient of numerous awards for teaching and research that include the NSF Presidential Young Investigator Award, 1988; the Bodossaki Foundation Award in Applied Sciences, 1997; the Aspen Tech Excellence in Teaching Award, 1999; and the 2001 AIChE Professional Progress Award for Outstanding Progress in Chemical Engineering.

Richard A. Friesner is professor of chemistry at Columbia University. He received his B.S. degree in chemistry from the University of Chicago and his Ph.D. from the University of California, Berkeley. Following postdoctoral work at the Massachusetts Institute of Technology, he joined the Chemistry Department at the University of Texas at Austin before moving to Columbia in 1990. His research, involving both analytical and computational theory, includes the application of quantum chemical methods to biological systems, development of molecular mechanics force fields and models for continuum solvation, computational methods for protein folding and structural refinement, prediction of protein-ligand binding affinities, and calculation of electron transfer rates in complex molecules and materials.

James R. Heath is the Elizabeth W. Gilloon Professor of Chemistry at the California Institute of Technology and he is the director of the California NanoSystems Institute, which was formed by California Governor Grey Davis in December 2000. Until 2003, he was professor of chemistry and biochemistry at the University of California, Los Angeles (UCLA). He received a B.Sc. degree in chemistry from Baylor University and a Ph.D. degree in chemistry from Rice University. Following postdoctoral work at the University of California, Berkeley, he was a research staff member at the IBM T.J. Watson Research Laboratories in Yorktown Heights, New York, from 1991 until 1994 when he joined the UCLA faculty. Heath's research interests focus on "artificial" quantum dot solids and quantum phase transitions in those solids; molecular electronics architecture, devices, and circuitry; and the spectroscopy and imaging of transmembrane proteins in physiological environments. He is a fellow of the American Physical Society and has received the Jules Springer Award in Applied Physics (2000), the Feynman Prize (2000), and the Sackler Prize in the Physical Sciences (2001).

Dimitrios Maroudas is Professor of chemical engineering at the University of Massachusetts, Amherst. Prior to accepting his present position in 2002, he was professor of chemical engineering at the University of California, Santa Barbara, and a visiting associate professor in the Department of Chemical Engineering at the Massachusetts Institute of Technology for the academic year 2000-2001. He graduated from the National Technical University of Athens with a diploma in chemical engineering, received his Ph.D. in chemical engineering with a minor in physics from the Massachusetts Institute of Technology, and did postdoctoral research at IBM's T.J. Watson Research Center, Yorktown Heights, New York. Professor Maroudas' research interests are in the area of theoretical and computational materials science and engineering. His research aims at the predictive modeling of structure, properties, dynamics, processing, and reliability of electronic and structural materials, especially semiconductor and metallic thin films and nanostructures used in the fabrication of electronic, optoelectronic, and photovoltaic devices. He has received a Faculty Early Career Development (CAREER) Award from the National Science Foundation, a Camille Dreyfus Teacher-Scholar Award, and several teaching awards.

Linda R. Petzold is professor in the Departments of Mechanical and Environmental Engineering, and Computer Science, and Director of the Computational Science and Engineering Program at the University of California, Santa Barbara. From 1978 to 1985 she was a member of the Applied Mathematics Group at Sandia National Laboratories in Livermore, California, and from 1985 to 1991 she was group leader of the Numerical Mathematics Group at Lawrence Livermore National Laboratory. From 1991 to 1997 she was professor in the Department of Computer Science at the University of Minnesota. She received her Ph.D. in computer science in 1978 from the University of Illinois. Her research interests include numerical ordinary differential equations, differential-algebraic equations, and partial differential equations, discrete stochastic systems, sensitivity analysis, model reduction, parameter estimation and optimal control for dynamical systems, multiscale simulation, scientific computing, and problem-solving environments. Dr. Petzold was awarded the Wilkinson Prize for Numerical Software in 1991 and the Dahlquist Prize for numerical solution of differential equations in 1999.

George C. Schatz is professor of chemistry at Northwestern University. He received a B.S. degree from Clarkson University and a Ph.D. in chemistry from California Institute of Technology in 1976. His research is aimed at understanding the interaction of light with nanoparticles and with nanoparticle aggregates. He is also actively working on structural modeling of DNA melting and of molecular self assembly processes. In addition, he is interested in time-dependent chemical processes such as bimolecular reactions and collisional energy transfer. He is a fellow of the American Physical Society, the American Association for the Advancement of Science, the International Academy of Quantum Molecular Science, and the American Academy of Arts and Sciences. He is a recipient of the Max Planck Research Award and serves as senior editor of the *Journal of Physical Chemistry*.

Larry L. Smarr is the Harry E. Gruber Professor of Computer Science and Information Technologies at the Jacobs School's Department of Computer Science and Engineering at the University of California, San Diego (UCSD). He is the founding institute director of the California Institute for Telecommunications and Information Technology, which brings together more than 200 faculty from UCSD and the University of California, Irvine and more than 50 industrial partners to research the future development of the Internet. Prior to moving to UCSD in 2000, he was on the faculty of the University of Illinois at Urbana-Champaign Departments of Physics and of Astronomy, where he conducted observational, theoretical, and computational-based research in relativistic astrophysics. In 1985 he was named the founding director of the National Center for Supercomputing Applications (NCSA) at the University of Illinois, and in 1997, he also became the director of the National Computational Science Alliance, comprised of more than 50 universities, government labs, and corporations linked with NCSA in a national-scale virtual enterprise. Smarr was an undergraduate at the University of

Missouri; he earned a master's at Stanford University and his doctorate from the University of Texas at Austin. He did postdoctoral work at Princeton University and was a junior fellow in the Harvard University Society of Fellows. He is a fellow of the American Physical Society and the American Academy of Arts and Sciences, and he received the 1990 Delmer S. Fahrney Gold Medal for Leadership in Science or Technology from the Franklin Institute. Dr. Smarr is a member of the National Academy of Engineering.

Ellen B. Stechel is manager of the Emissions Compliance Engineering Department for Ford Motor Company, North America Engineering, where she recently served as implementation manager for Ford's Accelerated Catalyst Cost Reduction Opportunity Program. Formerly, she was manager of the Chemistry and Environmental Science Department in the Scientific Research Laboratories at Ford Motor Company She received her A.B. in mathematics and chemistry from Oberlin College, her M.S. in physical chemistry from the University of Chicago, and her Ph.D. in chemical physics also from the University of Chicago. After a postdoctoral research position at UCLA, she joined the technical staff at Sandia National Laboratories, becoming manager of the Advanced Materials and Device Sciences Department in 1994. She left Sandia to accept a position at Ford Motor Company in 1998. She has served as a senior editor for the *Journal of Physical Chemistry* and has been active in several professional societies, including the American Vacuum Society (where she is currently a trustee).

Dennis J. Underwood is Vice President for discovery informatics and computational sciences at Infinity Pharmaceuticals, Inc., in Cambridge, Massachusetts. Before assuming his current position in 2003, he was a Director at Bristol-Myers Squibb, Wilmington (formerly DuPont Pharmaceuticals). He received his B.Sc. degree and his Ph.D. degree in physical organic chemistry at Adelaide University. Following postdoctoral work at Rockefeller University and Cornell University, he returned to Australia for additional postdoctoral work at the Commonwealth Scientific and Industrial Research Organisation (CSIRO) Division of Applied Organic Chemistry and then at the Australian National University. In 1985 he joined the molecular modeling group at the Merck Research Laboratories where he remained as head of molecular modeling until 1998. During this time, his research included the structure-based design of human leukocyte elastase inhibitors, virtual screening methodologies and studies of G-protein coupled receptors. Dr. Underwood moved to DuPont Pharmaceuticals as a senior director in 1998 and was responsible for both discovery informatics and the molecular design group. In his current position, he is responsible for structural biology (protein crystallography) and molecular design. In this role he has continued with development of novel computer methodologies aimed at drug discovery and development and has continued his work on structure-based design on G-protein coupled receptors as drug targets.

F

Participants

CHALLENGES FOR THE CHEMICAL SCIENCES IN THE 21ST CENTURY: WORKSHOP ON INFORMATION AND COMMUNICATIONS
October 31-November 2, 2002

Richard C. Alkire, University of Illinois at Urbana
Daniel Auerbach, IBM Research
Paul I. Barton, Massachusetts Institute of Technology
Charles H. Bennett, IBM Research Corporation
David L. Beveridge, Wesleyan University
Curt M. Breneman, Rensselaer Polytechnic Institute
Ronald Breslow, Columbia University
Donald M. Burland, National Science Foundation
Stanley K. Burt, National Cancer Institute
Anne M. Chaka, National Institute of Standards and Technology
Dennis I. Chamot, The National Academies
Thomas W. Chapman, National Science Foundation
Hongda Chen, U.S. Department of Agriculture
Zhong-Ying Chen, SAIC
Dean A. Cole, U.S. Department of Energy
Peter T. Cummings, Vanderbilt University
David A. Daniel, American Chemical Society
Juan J. de Pablo, University of Wisconsin, Madison
Brett I. Dunlap, U.S. Naval Research Laboratory
Thom H. Dunning, Jr., Joint Institute for Computational Sciences, University of Tennessee, Oak Ridge National Laboratory

Gregory Farber, National Institutes of Health
Christodoulos A. Floudas, Princeton University
Richard Friesner, Columbia University
Michael Frisch, Gaussian, Inc.
Christos Georgakis, Polytechnic University
Stephen Gray, Argonne National Laboratory
Mihal E. Gross, RAND, Science and Technology Policy Institute
Ignacio E. Grossmann, Carnegie Mellon University
Peter Gund, IBM Life Sciences
Carol Handwerker, National Institute of Standards and Technology
William L. Hase, Wayne State University
James R. Heath, University of California, Los Angeles
Judith C. Hempel, University of Texas at Austin
Richard L. Hilderbrandt, National Science Foundation
Daniel Hitchcock, U.S. Department of Energy
Peter C. Jurs, Pennsylvania State University
Miklos Kertesz, Georgetown University
Atsuo Kuki, Pfizer Global Research & Development
Michael Kurz, Illinois State University
Angelo Lamola, Worcester, Pennsylvania
Ku-Yen Li, Lamar University
Kenneth B. Lipkowitz, North Dakota State University
Mehrdad Lordgooei, Drexel University
Andrew J. Lovinger, National Science Foundation
Peter Ludovice, Georgia Institute of Technology
Dimitrios Maroudas, University of Massachusetts, Amherst
Glenn J. Martyna, IBM T.J. Watson Research Laboratory
Paul H. Maupin, U.S. Department of Energy
Gregory J. McRae, Massachusetts Institute of Technology
Julio M. Ottino, Northwestern University
Dimitrios V. Papvassiliou, University of Oklahoma
Linda R. Petzold, University of California, Santa Barbara
John A. Pople, Northwestern University
Larry A. Rahn, Sandia National Laboratory
Mark A. Ratner, Northwestern University
Robert H. Rich, American Chemical Society, Petroleum Research Fund
Celeste M. Rohlfing, National Science Foundation
David Rothman, Dow Chemical Company
George C. Schatz, Northwestern University
Peter P. Schmidt, Office of Naval Research
Michael Schrage, Massachusetts Institute of Technology
Robert W. Shaw, U.S. Army Research Office
Larry L. Smarr, University of California, San Diego

Ellen B. Stechel, Ford Motor Company
Walter J. Stevens, U.S. Department of Energy
Terry R. Stouch, Bristol-Myers Squibb Company
Michael R. Thompson, Pacific Northwest National Laboratory
M. Silvina Tomassone, Rutgers University
Mark E. Tuckerman, New York University
John C. Tully, Yale University
Dennis J. Underwood, Bristol-Myers Squibb Company
Ernesto Vera, Dow Chemical Company
Dion Vlachos, University of Delaware
Alfons Weber, National Science Foundation
Philip R. Westmoreland, University of Massachusetts
Ralph A. Wheeler, University of Oklahoma
William T. Winter, State University of New York

G

Reports from Breakout Session Groups

A key component of the Workshop on Information and Communications was the set of four breakout sessions that enabled individual input by workshop participants on the four themes of the workshop: discovery, interfaces, challenges, and infrastructure. Each breakout session was guided by a facilitator and by the expertise of the individuals as well as the content of the plenary sessions (Table G-1). Participants were assigned one of three groups on a random basis, although individuals from the same institution were assigned to different breakout groups. Each breakout group (color-coded as red, yellow, and green) was asked to address the same set of questions and provide answers to the questions, including prioritization of the voting to determine which topics the group concluded were most important. After every breakout session, each group reported the results of its discussion in plenary session.

The committee has attempted in this report to integrate the information gathered in the breakout sessions and to use it as the basis for the findings contained herein. When the breakout groups reported votes for prioritizing their conclusions, the votes are shown parenthetically in this section.

TABLE G-1 Organization of Breakout Sessions.

Breakout Session	Group	Facilitator	Session Chair	Rapporteur
Discovery	Red	D. Raber	J. Ottino	P. Jurs
	Yellow	P. Cummings J. Hempel	J. Tully	E. Vera
	Green	J. Jackiw	K. Lipkowitz	C. Breneman
Interfaces	Red	P. Cummings J. Hempel	J. Tully	G. McRea
	Yellow	J. Jackiw	K. Lipkowitz	R. Friesner
	Green	D. Raber	J. Ottino	P. Westmoreland
Challenges	Red	J. Jackiw	K. Lipkowitz	G. McRea
	Yellow	D. Raber	J. Ottino	L. Rahn G. Martyna
	Green	P. Cummings J. Hempel	J. Tully	P. Westmoreland
Infrastructure	Yellow-Red	D. Raber	J. Tully	M. Tuckerman (with G. McRea)
	Green-Red	J. Jackiw	K. Lipkowitz J. Ottino	P. Gund C. Breneman

DISCOVERY

What major discoveries or advances related to information and communications have been made in the chemical sciences during the last several decades?

Red Group Report

Prioritized List:

Analytical instrumentation and data visualization (11)
Computational materials science (10)
Process design and optimization (6)
Drug design and bioinformatics (5)
Environmental chemistry and modeling (5)
Materials applications (4)
Advances in quantum mechanics (3)

Yellow Group Report

Molecular Simulation of Chemical Systems

Materials, polymers, advanced materials, crystals and nanostructures, simulation techniques (DFT, Monte Carlo and MD)

Microfabrication by Chemical Methods (e.g., lithography)

Synthesis, design, and processing of new materials

Computer-Aided Drug Design

QSAR, molecular modeling

Numerical Simulation of Complex Chemical Processes Involving Reaction Transport

Ozone layer, combustion, atmospheric chemistry, chemical vapor deposition

Process Simulation, Optimization, and Control

Supply-chain optimization, plant design, and remote control of plants

Green Group Report

Electronic Structure Theory (9)

Common language of electronic structure theory, validating density functional theory, standardization of Gaussian basis sets, semiempirical methods, B3LYP functional (Becke's three-parameter hybrid functional using the LYP correlation functional), well-defined procedures for QM algorithms

Research Simulation Software for All Science (8)

Gaussian software, archiving chemical information, software development for the community, process control and process design software

Computer-Aided Molecular Design (7)

Combinatorial chemistry for drug study

Chemical Kinetics and Dynamics (5)

Coupling electronic structure theory with dynamics (direct dynamics), kinetics from first-principles—rate constants, stochastic simulation theory

Potential Functions and Sampling Methods (4)

Monte Carlo and molecular dynamics for statistical thermodynamics, potential energy functions and applications, discovery of applicability of potential energy functions, reactive empirical bond order potentials

Data-Driven Screening (2)

Application of statistical learning theory and QSPR, data mining and combinatorial high-throughput methods, virtual high-throughput screening

Synthesis of Materials for Computing (1)

Photolithography polymers, metallization, optical fibers

INTERFACES

What are the major computing-related discoveries and challenges at the interfaces between chemistry–chemical engineering and other disciplines, including biology, environmental science, information science, materials science, and physics?

Red Group Report

Targeted Design

How do we go backward from the desired specifications and functions to identify the molecules and processes?

A Matrix View Is Needed

Rows (discovery opportunities): biosytems modeling, proteins, drug design, materials design, environmental modeling (life-cycle analysis), molecular electronics, nanosciences

Columns (challenges to achieving discoveries): combinatorial search algorithms, data assimilation, multiscale modeling, structure-function and structure-activity relationships, large-scale computation, how to create interdisciplinary teams *as well as* support the critical need for disciplinary research

Results: the matrix is dense (every cell is a yes); people issues are critical; opportunities exist for both methodology development and application; there are many rows and columns, so the challenges are prioritization and maximizing the intersections; we need to show relevance, examples and benefits of what we are trying to do.

Yellow Group Report

Molecular Design (8)

Modeling at an atomic scale, small molecules vs. larger structures (one level of multiscale modeling); examples include design of drugs, nanostructures, catalysts, etc.

Materials Design (18)

More complex structures, additional length scales, property prediction (e.g., microelectronics devices, microfabrication, polymers, drug delivery systems, chemical sensors)

Multiscale Modeling (17)

Incorporates all length scales, important in all disciplines; examples include modeling of a cell, process engineering, climate modeling, brain modeling, environment modeling, combustion

Data Analysis and Visualization (6)

Analysis of high throughput data, seamlessness of large datasets, modeling of high-throughput datasets, information extraction, component architectures for visualization and computing

New Methods Development (6)

Green Group Report

Computer Science and Applied Math

Human-computer interface, expert systems, software engineering, high-performance computing, databases, visualization

Environmental Science

Fate of chemicals, computational toxicology, climate change, speciation

Biology

Protein structure-function; protein sequence-structure; networks (kinetic, regulatory, signaling); bioinformatics; dynamics in solvent environment

Physics

Chemical dynamics, scattering theory, molecular astrophysics, high-temperature superconductors

Materials Science and Engineering

Polymer property prediction, large-systems modeling (mesoscale simulation), nanostructures, corrosion

Earth Science

Water chemistry; binding of chemical species to ill-defined substrates; high-temperature, high-pressure chemistry; atmospheric chemistry

Crosscutting Issues

The issues are pervasive: every matrix entry is a yes.

Matrix columns (computing-related problem areas): software and interfaces (expert systems, development and maintenance and incentives); systems biology (chemistry, physics, and processes—complexity); molecular dynamics with quantitative reactive potentials; chemical and physical environment of calculations (e.g., solvation); simulation-driven experiments; special-purpose computing hardware; closing the loop for computational chemistry with analytical instruments and methods; database access, assessment, and traceability

Matrix rows (disciplines that interact with chemistry–chemical engineering): computational science and mathematics; environmental science; biology; physics; materials science and engineering; earth sciences

CHALLENGES

What are the information and communications grand challenges in the chemical sciences and engineering?

APPENDIX G

Red Group Report

Key Message

Reducing the time to solution

Some General Points

There are many possible information and communications challenges; *society:* no chemicals in this product; *education:* communicating excitement; *technical:* information gathering and exchange; new disciplines can emerge or have emerged from taking a systematic approach to knowledge storage and representation (e.g., genomics); if we are to be successful in our recommendations we need to address the challenges, opportunities and benefits (particularly with examples).

A Matrix View

Rows: grand challenge problems; *columns:* information and knowledge-based methodologies

Discovery Opportunities (Rows)

Simulating complex systems, cell, combustion, polyelectrolytes and complex fluids, atmospheres, hydrogeology, catalysts under industrial conditions, drug design, protein-DNA interactions, protein-protein, RNA folding, protein folding, metabolic networks, conformational sampling

Knowledge Issues

Knowledge database and intelligent agents; assimilating sensor–measurement–information data, data exchange standards (XML, etc.); representation of uncertainties in databases, building and supporting physical property archives (NIST, Webbook, etc.); collaborative environments

Methodology Issues

Model-based experimental design, long time scale dynamics, virtual measurements, force field validation and robustness, quantum dynamics, dispersion, excited states

Systems Approaches to Complex Problems

Intelligent and distributed databases (the "chemistry Google"), next-generation "laboratory notebooks" for models, databases, programs, and experimental data, next-generation databases, communication, sampling, and interaction across scales; *this is not a new issue.*

YELLOW GROUP REPORT

Long-Term Challenges

Comprehensive data and computational problem-solving environment (12): seamless chemical information, computer human interface

Simultaneous product and process development (12): for example, drug discovery, cradle-to-grave chemical production

Design of new molecule given required functionality (11): design lowest-energy consumption path, proof of concept for computational drug design

Virtual cell (4)

Design biocompatible materials (3): artificial tissue and bone, artificial organ, posttranslational protein modification

Short Term Challenges

Design-controlled directed self-assembly (10)
Self-assembly of protein in membrane (5)
Chemical plant on a chip (4)
Predict crystal structures (small organics) (4)
Computational electrochemistry (4)
Accurate free-energy calculation of medium size molecules (3)

Green Group Report

Molecularly Based Modeling

Electronic structure methods for 10^{4+} atoms, binding energies in arbitrarily complex materials, molecular processes in complex environments, computational tools to study chemical reactions (first principles, limits of statistical reaction rate theories), simulators with quantum dynamics for quantum degrees of freedom

Systems Issues

Integration of applied math and computer science to obtain more powerful software tools for the chemical sciences and engineering, modeling the origin and development of life, alternative chemical bases for life (other worlds), computational disease modeling, design of bio-inspired materials, global climate control (the earth as a chemical factory—similar to Groves' chip as a chemical factory), modeling of atmospheric chemistry

Measurements

Incorporation of computational chemistry methods into instruments, virtual measurement, enabling ubiquitous sensing, in vivo biosensors

Education and Usage Issues

Translate molecularly based modeling into education, expert systems for method choices (precision in context of time and cost)

INFRASTRUCTURE

What are the issues at the intersection of computing and the chemical sciences for which there are structural challenges and opportunities—in teaching, research, equipment, codes and software, facilities, and personnel?

Yellow-Red Group Report

Topics To Be Addressed

Teaching, research, equipment, codes and software, facilities, personnel

Teaching-Related Infrastructure Issues

Computational chemistry courses, interdisciplinary computational science courses, software engineering (not just programming), test problems and design case studies, support for software maintenance, modeling- and simulation-based courses; *top two issues:* modeling-and-simulation course, and mathematics training

Research-Related Infrastructure Issues

Multiscale modeling algorithms, representation and treatment of uncertainties, standard test cases (software, experiments), funding of interdisciplinary research, shared instruments; *top two issues:* multiscale modeling algorithms and funding of interdisciplinary research

Equipment-Related Infrastructure Issues

Access to high-performance computational facilities; high-bandwidth access to instruments and computers; collaborative environments; clusters and tightly coupled, diverse architectures; assimilation of information from sensors; interoperability and portability; distributed databases; shared resources (equipment, software, computers); access to a diverse set of architectures; *top two issues:* access to a diverse set of architectures, and interoperability and portability

Codes and Software-Related Infrastructure Issues

Integration of multivendor software systems, software maintenance (laboratory technicians), open source, code and data warehouses, component architectures, interoperability, security; *top two issues:* component architectures and open source code

Personnel Related-Infrastructure Issues

Poor computer and software engineering skill levels of incoming students, tenure and promotion of interdisciplinary people, training of chemists with chemical engineering concepts, people in the pipeline, attracting interest of computer science community; *top two issues:* tenure and promotion of inter-disciplinary people, and people in the pipeline

Green-Red Group Report

What Parts Are Working?

Commercial software companies serving this area: computational chemistry, chemical processes
Chemist–chemical engineer collaboration works well where it exists
Modern programming tools have speeded code development
Networking: Internet high-speed connectivity

What Parts Are Not Working?

Commercial software efforts limited by market size: limited new science, porting to new machines, commercialization can kill technology, coding complex programs still very difficult, high-performance computing centers are not generally useful, proposal process too restrictive, DOE centers working better than NSF centers
Education of new computational chemists and chemical engineers inadequate: undergraduate–graduate curriculum needs to address these subjects, applications and scientific programming (Colorado School of Mines Engineering Web site is doing this)
Academic code sharing and support mechanisms poor: code development

not supported by grants, no money or staff for maintenance and support, duplication of effort on trivial code; need open source toolkits and libraries

Databases and knowledge bases: not generally available except in bioinformatics, need standards and data validation

Cross-discipline collaboration and teamwork inadequate: overspecialization in graduate school departments

What Payoffs Are Expected?

Proof of profit—demonstration that simulation methods can make difference in business outcome, "home run" in drug discovery, more incorporation of methods into educational programs (better understanding of scope and limitations of methods), make specialists in industry more connected and useful (leveraging current staff), better communication and teaming of chemists and chemical engineers (more joint appointments, maybe combining departments), commonality in understanding and language between users and nonusers

How Can Community Assure Reliability, Availability, and Maintenance of Codes?

Recognize contributions of coders, support full life cycle of codes, agencies should fund code development and maintenance as part of grants, demonstrate value of simulation codes, funding incentives for cross-disciplinary software creation

H

List of Acronyms

ACS	American Chemical Society
ADME-Tox	absorption, distribution, metabolism, excretion, and toxicity
AFOSR	Air Force Office of Scientific Research
AFS	Andrew File System
AIChE	American Institute of Chemical Engineers
ASCI	Advanced Simulation and Computing Program
BOMD	conventional Born-Oppenheimer molecular dynamics
CBS	complete basis set
CCCBDB	NIST Computational Chemistry Comparison and Benchmark Database
CCSD	coupled cluster theory with single and double excitations
CCSD(T)	CCSD with perturbative correction for triple excitations
CI	configuration interaction
CMDE	Collaborative Modeling-Data Environments
CPMD	Car-Parinello molecular dynamics
DAE	differential-algebraic equations
DD	direct dynamics
DFT	density functional theory
ECEPP/3	Empirical Conformational Energy Program for Peptides
EMSL	Environmental Molecular Sciences Laboratory

APPENDIX H

GB	generalized Born
HF	Hartree-Fock
ILP	Integer Linear Programming
IT	information technology
G2	Gaussian-2 theory
LD	linkage disequilibrium
LDA	local density approximation
LLNL	Lawrence Livermore National Laboratory
MD	molecular dynamics
MP2	second order perturbation theory
MRI	magnetic resonance imaging
ODE	ordinary differential equations
QM-MM	quantum mechanics–molecular mechanics
MEMS	microelectromechanical system(s)
NASA	National Aeronautics and Space Administration
NCREN	North Carolina Research and Education Network
NCSA	National Center for Supercomputing Applications
NFS	Network File System
NIH	National Institutes of Health
NIST	National Institute of Standards and Technology
NMR	nuclear magnetic resonance
NSEC	National Science and Engineering Center
NSF	National Science Foundation
PDB	Protein Data Bank
PDE	partial differential equations
PB	Poisson-Boltzmann
PD	pattern discovery
PETSc	Portable, Extensible Toolkit for Scientific Computing
PNNL	Pacific Northwest National Laboratory
PSC	Pittsburgh Supercomputing Center
PUNCH	Purdue University Network Computing Hubs
QCD	Quantum ChromoDynamics
QCDOC	QCD on a chip

QM-MM	quantum mechanics–molecular mechanics
QSAR	quantitative structure-activity relationship
QSPR	quantitative structure-property relationship
SNP	single nucleotide polymorphisms
UNICORE	Uniform Interface to Computing Resources
VNMRF	Virtual NMR Facility
WTEC	World Technology Evaluation Center, Inc.